BIOTECHNOLOGY BY OPEN LEARNING

Infrastructure and Activities of Cells

PUBLISHED ON BEHALF OF :

Open universiteit   and   **Thames Polytechnic**

Valkenburgerweg 167   Avery Hill Road
6401 DL Heerlen   Eltham, London SE9 2HB
Nederland   United Kingdom

BUTTERWORTH
HEINEMANN

Butterworth-Heinemann Ltd
Linacre House, Jordan Hill, Oxford OX2 8DP

 PART OF REED INTERNATIONAL BOOKS

OXFORD   LONDON   BOSTON
MUNICH   NEW DELHI   SINGAPORE   SYDNEY
TOKYO   TORONTO   WELLINGTON

First published 1991

British Library Cataloguing in Publication Data
A catalogue record for this book is
available from the British Library

Library of Congress Cataloguing in Publication Data
A catalogue record for this book is
available from the Library of Congress

ISBN 0 7506 1500 1

Composition by Thames Polytechnic
Printed and bound in Great Britain by Thomson Litho, East Kilbride, Scotland

The Biotol Project

The BIOTOL team

OPEN UNIVERSITEIT, NETHERLANDS
Dr M. C. E. van Dam-Mieras
Professor W. H. de Jeu
Professor J. de Vries

THAMES POLYTECHNIC, UK
Professor B. R. Currell
Dr J. W. James
Dr C. K. Leach
Mr R. A. Patmore

This series of books has been developed through a collaboration between the Open universiteit of the Netherlands and Thames Polytechnic to provide a whole library of advanced level flexible learning materials including books, computer and video programmes. The series will be of particular value to those working in the chemical, pharmaceutical, health care, food and drinks, agriculture, and environmental, manufacturing and service industries. These industries will be increasingly faced with training problems as the use of biologically based techniques replaces or enhances chemical ones or indeed allows the development of products previously impossible.

The BIOTOL books may be studied privately, but specifically they provide a cost-effective major resource for in-house company training and are the basis for a wider range of courses (open, distance or traditional) from universities which, with practical and tutorial support, lead to recognised qualifications. There is a developing network of institutions throughout Europe to offer tutorial and practical support and courses based on BIOTOL both for those newly entering the field of biotechnology and for graduates looking for more advanced training. BIOTOL is for any one wishing to know about and use the principles and techniques of modern biotechnology whether they are technicians needing further education, new graduates wishing to extend their knowledge, mature staff faced with changing work or a new career, managers unfamiliar with the new technology or those returning to work after a career break.

Our learning texts, written in an informal and friendly style, embody the best characteristics of both open and distance learning to provide a flexible resource for individuals, training organisations, polytechnics and universities, and professional bodies. The content of each book has been carefully worked out between teachers and industry to lead students through a programme of work so that they may achieve clearly stated learning objectives. There are activities and exercises throughout the books, and self assessment questions that allow students to check their own progress and receive any necessary remedial help.

The books, within the series, are modular allowing students to select their own entry point depending on their knowledge and previous experience. These texts therefore remove the necessity for students to attend institution based lectures at specific times and places, bringing a new freedom to study their chosen subject at the time they need it and a pace and place to suit them. This same freedom is highly beneficial to industry since staff can receive training without spending significant periods away from the workplace attending lectures and courses, and without altering work patterns.

Contributors

AUTHORS

Dr C. K. Leach, Leicester Polytechnic, Leicester, UK

Dr J. Sampson, University of Leicester, Leicester, UK

Dr G. D. Weston, Leicester Polytechnic, Leicester, UK

EDITOR

Dr C. K. Leach, Leicester University, Leicester, UK

SCIENTIFIC AND COURSE ADVISORS

Dr M. C. E. van Dam-Mieras, Open universiteit, Heerlen, The Netherlands

Dr C. K. Leach, Leicester Polytechnic, Leicester, UK

ACKNOWLEDGEMENTS

Grateful thanks are extended, not only to the authors, editors and course advisors, but to all those who have contributed to the development and production of this book. They include Dr N. Chadwick, Dr G. Lawrence, Miss J. Skelton, Professor R. Spier and Mrs M. Wyatt. Special thanks go to Dr M. Walker and her colleagues (University of Leicester) for the electron micrographs used in this text. The development of this BIOTOL text has been funded by COMETT, The European Community Action programme for Education and Training for Technolgy, by the Open universiteit of The Netherlands and by Thames Polytechnic. Thanks are also due to the authors and editors of Open universiteit materials upon which some of this text was based.

Contents

How to use an open learning text

An open learning text presents to you a very carefully thought out programme of study to achieve stated learning objectives, just as a lecturer does. Rather than just listening to a lecture once, and trying to make notes at the same time, you can with a BIOTOL text study it at your own pace, go back over bits you are unsure about and study wherever you choose. Of great importance are the self assessment questions (SAQs) which challenge your understanding and progress and the responses which provide some help if you have had difficulty. These SAQs are carefully thought out to check that you are indeed achieving the set objectives and therefore are a very important part of your study. Every so often in the text you will find the symbol Π , our open door to learning, which indicates an activity for you to do. You will probably find that this participation is a great help to learning so it is important not to skip it.

Whilst you can, as a open learner, study where and when you want, do try to find a place where you can work without disturbance. Most students aim to study a certain number of hours each day or each weekend. If you decide to study for several hours at once, take short breaks of five to ten minutes regularly as it helps to maintain a higher level of overall concentration.

Before you begin a detailed reading of the text, familiarise yourself with the general layout of the material. Have a look at the contents of the various chapters and flip through the pages to get a general impression of the way the subject is dealt with. Forget the old taboo of not writing in books. There is room for your comments, notes and answers; use it and make the book your own personal study record for future revision and reference.

At intervals you will find a summary and list of objectives. The summary will emphasise the important points covered by the material that you have read and the objectives will give you a check list of the things you should then be able to achieve. There are notes in the left hand margin, to help orientate you and emphasise new and important messages.

BIOTOL will be used by universities, polytechnics and colleges as well as industrial training organisations and professional bodies. The texts will form a basis for flexible courses of all types leading to certificates, diplomas and degrees often through credit accumulation and transfer arrangements. In future there will be additional resources available including videos and computer based training programmes.

Preface

This is the second of two BIOTOL texts focusing on cells as the basic operational units of biological systems. The underpinning theme of these texts is to consider cells as biological 'factories' in so far as they take in one set of chemicals (nutrients) and convert them into new products.

Cells are quite superb 'factories'. Each cell type taking in its own set of chemicals and making its own collection of products. The products are themselves quite remarkable, ranging from the chemically simple, such as ethanol, to the complex, such as antibodies and hormones.

To understand how a factory operates requires knowledge of the tools and equipment available within the factory and how the use of these resources is managed and co-ordinated to provide an effective unit. Understanding the properties and activities of cells clearly necessitates knowledge of the structure and properties of the building blocks that make up cells and the nature of the 'tools' used within cells. This aspect of cell biology is the topic of the BIOTOL text 'The Molecular Fabric of Cells'. For a factory to operate effectively, however, it is not just a matter of acquiring the right tools to carry out the job. It is also essential to organise and manage the use of these tools in a co-ordinated and functional manner. This text is primarily concerned with examining the ways in which the activities of cells are organised and co-ordinated.

There is however one special feature of biological 'factories', not characteristic of man-made factories namely the ability of one factory (cell) to reproduce itself thus forming many similar factories. Implicit in this self-propagation is the need to not only produce additional sets of tools and building blocks but also to ensure the new factories receive an appropriate management structure to ensure they work efficiently. A substantial part of this text is concerned with the issues of cell proliferation.

The reader should recognise that all the cells in the larger (multicellular) systems do not carrying out identical processes. Multicellular systems (particular plants and animals) produce a variety of cell types, each designed to fulfil particular functions (ie there is a 'division of labour'). Thus some might be responsible for absorbing chemicals from the environment, others might be involved in transport or in providing physical support or expelling waste material. Therefore, if we are to properly understand the activities of cells, we not only need knowledge of the basic properties of cells but also an appreciation of how cell specialisation is achieved. Such specialised (differentiated) cells however must be produced in the right numbers and in the right place within the organism. This text finishes with a discussion of how this co-operation between populations of cells is achieved. Although this text has been written with potential biotechnologists in mind, the subject area covered by this text will be of value in any area of applied biology.

Finally, this text has been written on the assumption that the reader is familiar with the major molecular species found in biological systems. Although this assumption has been made, extensive use of 'molecular reminders' have been given within this text. The reader is not, therefore, completely abandoned to their own memory of bio-molecules. For those without any, or with very limited background knowledge of biological chemistry, we would recommend the BIOTOL text 'The Molecular Fabric of Cells' which has been especially tailored to provide an appropriate knowledge upon which a firm understanding of cell biology can be built.

Scientific and Course Advisors: Dr M. C. E. van Dam-Mieras
Dr C. K. Leach

Project Manager: Dr J. W. James

The architecture of prokaryotic cells

The architecture of prokaryotic cells

1.1 Introduction

The idea that all living systems are made up of cells has a long history. The invention of the compound microscope by Jenson in 1590 and its development and use by Leeuwenhoek during the period 1650-1700 enabled biological material to be more closely examined than had hitherto been possible. Robert Hooke (1665) first used the term 'cells' to describe the box-like structure he observed in thin sections of plant materials. Subsequent observations especially by Turpin (1826) and Diyardin (1835) enabled Schleiden and Schwann (1838) to develop and enlarge a 'cell-theory' for living things. This cell theory has of course been further extended and refined over the intervening years but it is still central to our understanding of living systems. The basis of this theory is that all living things are constructed of units (cells) each of which develop from pre-existing units (cells).

cell wall

protoplasm

These early students also recognised that cells consisted of two main parts, an outer coat or cell wall and an inner 'gelly' - named by Purkinje (1859) as protoplasm. Although it was recognised that cells from plants and animals had some fundamental differences especially in terms of the thickness of the cell wall, it was generally accepted that all cells contained protoplasm and that this was the physical basis of all life.

nucleus

cytoplasm

The further development of light microscopes and staining procedures to highlight particular structures enabled a description to be made of sub-cellular components. Nucleus, plastids and cytoplasm became part of the biologists vocabulary.

prokaryotic

eukaryotic

The advent of the electron microscope earlier this century with a much greater magnifying power than that of light microscopes, led to greater resolution of the fine structure of cells. These studies clearly demonstrated that cells can be divided into two quite distinct types described as prokaryotic and eukaryotic. Prokaryotic cells are structurally the simpler of the two. This type of cell morphology is confined to the bacteria. The cells of other groups of micro-organisms (eg algae, fungi and protozoa) and of all plants and animals are of the eukaryotic type.

unicellular

multicellular

Prokaryotic cells are much smaller than eukaryotic cells. Most of those organisms which exhibit prokaryotic organisation are unicellular (ie each organism consists of a single cell) whilst many eukaryotes exhibit multicellularity (ie each organism consists of many cells which may be of many different types).

In this chapter, you will learn about the basic structure of prokaryotic cells. Subsequent chapters will deal with eukaryotic cell structure and function. Despite the fact that prokaryotic organisation is restricted to the prokaryotes, it is still important to have a good understanding of how such cells function since bacteria include some important disease causing types as well as potential mediators of many important biotechnological processes. They also include some types which are responsible for major environmental and geochemical changes. The fundamental differences between prokaryotes and eukaryotes has important consequences in medicine, genetic manipulation and biotechnology.

This first chapter is quite long so do not attempt to do it all in one sitting.

1.2 The diversity of prokaryotic cells

prokaryotes

eubacteria

archaebacteria

Although all prokaryotic cells are relatively simple structures, there are many subtle variations associated with their basic architecture. Here we will confine ourselves to the common and major features found amongst prokaryotes (prokaryotes = organisms which display prokaryotic cell organisation). We can, however, distinguish two sub-groups of prokaryotes - the true bacteria (eubacteria) and the archaebacteria.

If we examine the fine structure of the eubacteria and archaebacteria through a powerful microscope, they are remarkably similar to each other. If, on the other hand, we analysed the chemicals which make up these two types, then we would discover that they are quite different. For the most part, the chemicals which make up the eubacteria are very similar to those that make up plants and animals (eukaryotes). The archaebacteria, on the other hand, contain many chemicals which are quite distinct from those found in eubacteria, plants and animals.

The eubacteria include most of the bacteria that are commonly found in soil, water and on other living systems. Thus the bacteria which cause diseases, that breakdown plant debris in our gardens or are found in our rivers are usually of the eubacterial type (ie of chemical composition similar to that of plants and animals).

thermo-
acidophiles,
halophiles,
methanogens

In contrast, archaebacterial organisms are usually found in uncompromising environments such as hot acid springs (thermoacidophiles) or in salt waters and brines (halophiles) or are capable of generating methane from carbon dioxide (methanogens). Below is a checklist of the basic divisions of cell types (Table 1.1).

Cell Structure	Group	Properties	Examples
Prokaryotic	Eubacteria	simple structure chemically similar to eukaryotes	most bacteria, including disease organisms, green photosynthetic bacteria, cyanobacteria (blue-green algae), purple photsynthetic organisms
Prokaryotic	Archaebacteria	simple structure chemically quite different to eukaryotes	thermoacidophiles, halophiles, methanogens
Eukaryotic	Eukaryotes	can be unicellular or multicellular, basic cell architecture is much more complex than with prokaryotes	micro-organisms (algae, fungi, protozoa), plants (mosses, ferns, seed plants), animals (invertebrates, vertebrates)

Table 1.1 The major divisions of cell types.

For the remainder of this chapter, we will predominantly discuss the eubacteria.

1.3 The morphology and fine structure of the eubacteria

We can conveniently divide the discussion of the structure of eubacteria into:

- a consideration of the gross morphology of the cells including size, shape and arrangement;

- discussion of the fine structure of the cells.

Such a division more-or-less mimics the history of descriptive cell biology. Prior to the 1940's only light microscopes were available. Although such microscopes could be made powerful, giving magnification up to about x1000, all prokaryotic cells are very small and little more than their general shape, the arrangement of the cells and a few major structures such as spores could be distinguished.

The advent of the electron microscope in the 1940's provided greater magnification and hence much greater resolution thereby enabling much finer structure to be identified and described. These advances were also matched by improvements in cell fractionation and biochemical analytical techniques and thus we began to be able to relate observed structures to chemical composition and function.

1.3.1 Gross morphology of bacterial cells

Size

Bacteria are all very small. They are measured in micrometers (written μm). 1 μm is equivalent to 10^{-6} m or 10^{-3} mm. Typically bacteria have widths in the range of 0.5-2 μm. In other words, we could lay about 1000 bacteria side by side to cover about 1 mm.

2-5 μm long,
1 μm wide

Although bacteria have relatively uniform widths, they can vary considerably in their length. Most bacteria fall into the range of being 2-5 μm long - a few however, may be over 100 μm long.

mycoplasma

The smallest bacteria known belong to a group known as mycoplasma. Mycoplasma have a size range of 0.1 to 0.3 μm.

The small sizes of bacterial cells mean that they are close to the limits of resolution of a light microscope. In other words, they can only just be seen under the highest magnification. Let us do a simple calculation to show that this is true.

∏ The limit of resolution of the naked eye is about 0.2 mm. This means that if we put two dots on a piece of paper closer together than 0.2 mm, they will appear as a single spot. (Try it!). The most powerful light microscope has a magnifying power of approximately x1000. Using these two facts, what are the limits of resolution using such a microscope?

Using the microscope, objects (including the distances between objects) can be magnified by x1000. Thus by using such a microscope, we can effectively improve the resolving power of the eye by x1000. Thus if we put two dots 0.2/1000 mm apart under the microscope, we should be able to distinguish them as two separate dots since through the microscope they will appear to be 0.2 mm apart.

Through the microscope therefore our resolving power = 0.2/1000 mm = 0.2 µm. Ojects smaller than this will not be resolved as separate objects.

The sizes given for bacteria indicate that these cells are amongst the smallest objects that can be seen using a light microscope. The mycoplasmas are barely visible even with the very best light microscopes.

The small size of prokaryotic cells has many important consequences other than just being difficult to see. Of perhaps greatest importance is the ratio of the surface area to the volume of these cells. Let us again do a simple calculation.

$\prod$ Consider a cube 1 mm x 1 mm x 1 mm, its volume is 1 x 1 x 1 = 1 mm^3.

Its surface area is 1 x 1 x 6 = 6 mm^3 (6 faces each 1 mm x 1 mm).

The relative surface : volume ratio is therefore 6:1.

Now let us divide the cube into smaller cubes each with a length of 0.1 mm.

What will be the surface to volume ratio now? The total volume will remain the same (ie 1 mm^3) but the surface area will be 0.1 x 0.1 (area of one face) x 6 (number of faces per cube) x 1000 mm^2 (number of small cubes) = 60 mm^2. The relative surface : volume ratio is now 60:1.

$\prod$ Let us take this calculation one stage further. What would be the relative surface : volume ratio if we divided the cube into 1 µm (10^3 mm) long cubes (ie about the size of a bacterium)?

The total volume would remain at 1 mm^3. The area of each face of the small cubes would be 0.001 x 0.001 mm^2. Each cube would have 6 faces. There would be 1000 x 1000 x 1000 small cubes in each mm^3.

Therefore the total surface area = 0.001 x 0.001 x 6 x 1000 x 1000 x 1000 mm^2 = 6000.

The relative surface : volume ratio would be 6000:1.

high rate of metabolism What this calculation will have shown you is that prokaryotic cells which are very small, have extremely high surface : volume ratios. Much higher than for example that of the larger eukaryotic cells. The consequences of this are important for it means that the contents of prokaryotic cells have a large interface with the nutrients in the environment. This large area of interface allows for high rates of uptake of nutrients. This in turn allows for very high rates of metabolism and, subsequently, growth of the cells. It is because of this very small size that prokaryotic cells are the most efficient, in terms of speed of turnover, chemical 'factories' found in the biological world. Thus although each bacterial cell is small (perhaps weighing about 10^{-12} g), on a weight to weight basis, bacteria have metabolic turnover rates 10,000 - 1,000,000 times faster than, for example, large animals. These extremely fast rates of chemical turnover support very fast growth rates. Many prokaryotes can double their mass in 20-30 minutes. This, coupled with the diversity of metabolism displayed by these types of cells and the increasing ability to genetically modify them, makes these simple cells attractive for use in biotechnological processes.

Shape

Prokaryotic cells take up a limited number of general shapes (Figure 1.1). These are given special terms:

cocci
- spherical cells are called cocci (singular = coccus = berry);

bacilli
- cylindrical rods are called bacilli (singular = bacillus = rod);

spirillal cells
- spiral shaped cells are called spirillal cells;

vibrios
- part spiral or comma shaped cells are called vibrios.

Not all bacilli are exactly the same shape, some are long and thin (eg *Clostridium sporogenes*), others short and fat (*Bacillus megaterium*), some have square ends, others are tapered (NB descriptive microbiology includes many words to describe these fine details such as fusiform, ellipsoid etc). We can perhaps imagine a continuous spectrum of cell shapes as shown in Figure 1.1.

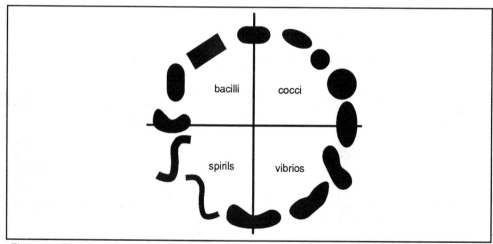

Figure 1.1 General shapes of prokaryotic cells.

pleomorphism
Some species of prokaryotes produce cells of more than one shape. Such cells are said to exhibit pleomorphism.

Arrangement of prokaryotic cells

The manner in which cells are arranged often reflect the way in which the cells grow and divide. Let us see if you can predict some of these arrangements. Consider a coccus which can grow and divide in *one* plane only.

If we start with one cell and this grows and divides into two and then four, how will the cells be arranged?

streptococci

We now have a short chain. Thus cocci, which grow and divide in only one plane, form chains. They are called streptococci. Streptococci are quite common, they include some of the organisms which are responsible for the souring of milk, while others are responsible for infections such as septic sore throats. With some cocci which divide in one plane only, the chain breaks up into double cells thus:

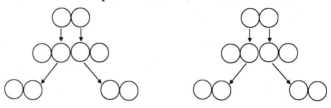

diplococci

Since these cells are usually found in twos they are called diplococci.

∏ Use a piece of paper to make some rough drawings to answer the following questions: 1) What would be the arrangement of cocci if the cells could divide in any plane? 2) What would be the arrangement of the bacilli, if each bacillus divided transversely across the cell and the progeny (daughter cells) failed to separate? 3) What would be the arrangement of the bacilli if each bacillus divided longitudinally?

You probably predicted that:

1) the cocci would be arranged in clusters. Thus:

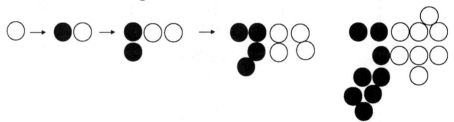

staphylococci

Such organisms are said to be staphylococci. Some important organisms that live on the surface of human skin are staphylococci.

2) the bacilli would be arranged into a long filament or chain.

bacillus

Many bacillal forms produce long chains (eg *Bacillus cereus*). There is however a high degree of variability as to the extent to which the progeny are held together after division and many bacillal forms exist as unicells.

3) Those bacilli that divide longitudinally will if the progeny (daughter cells), fail to separate, form a kind of palisade much like a row of matchsticks.

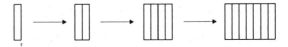

With many bacilli which divide longitudinally, division is also accompanied by the separation of the daughter cells.

Sometimes this is accomplished by a 'bending' division

or by a 'snapping' division.

Careful examination of cultures of prokaryotic cells can therefore provide many clues as to the manner in which the cells grow and divide.

Now attempt SAQ 1.1 before moving on to the next section.

1.3.2 Staining of prokaryotes

Now that we know something about the size, shape and arrangement we find amongst prokaryotic cells, let us turn our attention to another important aspect. We have learnt that prokaryotic cells are very small and are close to the limits of resolution of light microscopes. Their minute size has another consequence. They are so thin that they do not absorb much light (except for the photosynthetic types which deliberately produce pigments to absorb light). Since they do not absorb much light, it is difficult to see them under a microscope using bright field illumination. In order to increase the contrast between the cells and their background, it is usual to use dyes to stain them.

cationic dyes   Dyes are organic compounds which absorb specific wavelengths of light. They often contain ionisable groups (eg $-NH_3^+$) and carry a net electrostatic charge. Positively charged dyes combine with negatively charged cell constituents. Positively charged dyes are also known as basic dyes or cationic dyes.

∏   Make a list of cell constituents such positively charged dyes might combine with.

The most likely candidates are nucleic acids (RNA, DNA), acidic polysaccharides and acidic proteins. Nucleic acids carry negative charges on their phosphate groups at cellular pH's, acidic polysaccharides and proteins carry carboxylic acid groups which can dissociate to give a net negative charge.

Such molecules attract and hold cationic dyes. The cell walls of prokaryotes contain acidic polysaccharides and their cytoplasms contain considerable amounts of nucleic acids. Prokaryotic cells are therefore usually very readily stained with cationic dyes.

anionic dyes   Negatively charged (anionic or acidic) dyes combine with positively charged components such as basic proteins. Basic proteins contain high levels of the amino acids lysine and arginine, these have amine and imine groups which can carry a net positive charge thus:

$$-NH_2 \quad \rightarrow \quad N^+H_3 \text{ at cellular pH's}$$

$$= NH \quad \rightarrow \quad = N^+H_2$$

SAQ 1.1

Examine the following drawings of cultures of prokaryotic cells carefully.

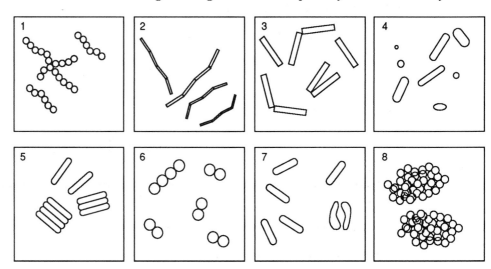

Answer the following questions concerning these cultures.

1) Which of the cultures are composed of cocci?

2) Which of the cultures contain cells which divide in a single plane?

3) Which one of the cultures can best be described as being composed of staphylococci?

4) Which of the cultures should be described as being pleomorphic?

5) Which of the cultures exhibit 'snapping' division?

6) Which of the cultures are composed of vibrios?

7) Which of the cultures can best be described as being composed of streptococci?

8) Which of the cultures is most likely to have the slowest growing cells?

hydrophobic dyes Dyes which do not carry a charge (eg Sudan Black) are hydrophobic and prefer to find a non-polar environment (they are fat soluble). Dyes like Sudan Black are useful for staining oil droplets and deposits of fats.

Let us examine the procedures adopted for staining prokaryotic cells. We can divide staining procedure into two types - simple staining and differential staining.

1.3.3 Simple staining

Simple staining is, as the name implies, relatively straightforward. The steps are:

• place a thin smear of the specimen on a glass slide;

- dry it carefully, then heat it strongly (but not too strongly so that it burns) to cause the cells to stick onto the glass slide;

- flood the smear with the stain, leave for a specified time, wash off with water and allow to dry;

- view.

With this type of procedure most cationic dyes will stain the cells almost uniformly. Some stains (eg Sudan Black) only stain specified structures such as oil storage droplets. Such simple staining procedures are used predominantly just to aid visualisation of the cells under the microscope. Stains can, however, be used to do more than just provide a contrast so that cells can be more easily seen. They can be used to help identify or differentiate (distinguish) between similar shaped organisms. This is the basis of differential staining.

differential staining

1.3.4 Differential staining

The most important and widely used differential staining procedure is the Gram stain. The sequence is given in Figure 1.2.

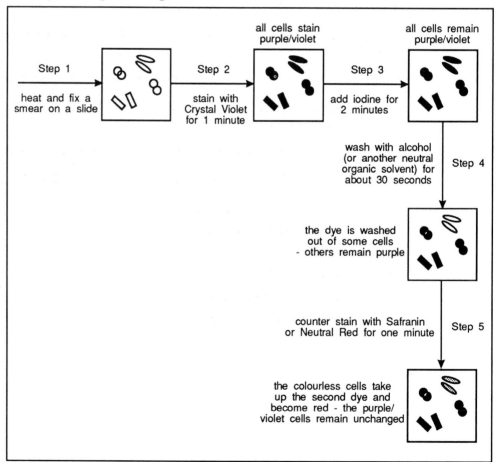

Figure 1.2 The stages of the Gram staining procedure.

Gram positive

Gram negative
If we examine what has happened to the cells in this sequence, some retained the first stain used (Crystal Violet). Such cells are said to be Gram positive. Those cells which lose the first stain when they are washed with alcohol are said to be Gram negative. Using this procedure we can therefore divide the cells into one of two types. They are either Gram positive or Gram negative.

∏ Can you think of some reasons why some cells retain the Gram stain while others do not?

The difference appears to be in the nature of the surface of the cell and is a reflection of quite fundamental differences. Although described as early as 1884 by Christian Gram, the technique is still widely used particularly in clinical circumstances as an aid to identification of certain bacteria. We will learn more of the fundamental differences between Gram positive and Gram negative organisms later in this chapter. For now, let us see if we can use the Gram's stain to generalise about differential staining.

The sequence used in Gram's staining procedure can be divided up into the following steps:

- primary stain (in this case Crystal Violet);

- fixative to 'fix' the primary stain onto or into the cells. In the case of the Gram stain, iodine is used;

- challenge (attempt to wash out the primary stain). In this case, alcohol is used;

- counterstain (to stain the cells which have been decolourised by the challenge). In this case Safranine or Neutral Red.

Try to keep the sequence 'primary stain, fixative, challenge and counterstain' in mind.

acid fast stain Here is another example of a differential stain. This is called the acid fast stain.

Step 1 - Heat fix a smear onto a slide

Step 2 - Flood with Carbol-Fuchsin

Step 3 - Heat over a steam bath for 10 minutes

Step 4 - Cool and flood with dilute acid, wash off

Step 5 - Stain with Methylene Blue

Now, attempt the following SAQ.

1) In the staining procedure described above as the acid fast stain:

What stain is used as the primary stain?

What procedure is used to 'fix' the primary stain?

What reagent is used as a challenge (ie to attempt to wash out the primary stain)?

What is used as the counter stain?

2) Why is this procedure called the acid fast stain?

3) What colour will non-acid fast cells appear?

We will meet some other staining procedures when we examine the fine structure of prokaryotic cells.

1.4 The fine structure of prokaryotic cells

Figure 1.3 represents a highly stylised section through a prokaryotic cell. The figure is provided with a scale so that the approximate sizes of the labelled structures can be estimated.

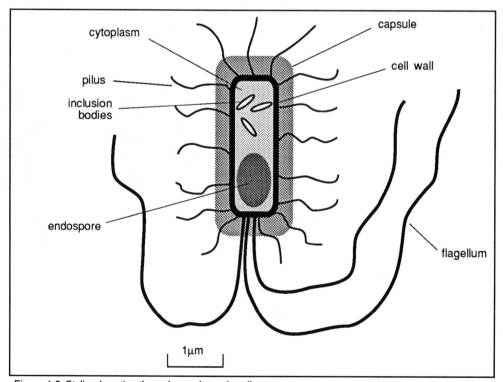

Figure 1.3 Stylised section through a prokaryotic cell.

SAQ 1.3

Using Figure 1.3 and its bar scale, circle the structures listed below which can normally be seen through a light microscope? (Remember our earlier calculation of the limits of resolution of light microscopes).

Cell wall, endospore, pilus, flagellum, inclusion bodies, capsule.

1.4.1 Capsules and slime layers

Many prokaryotes produce polymers that are deposited on the outside of the cell wall as a more-or-less homogenous layer. When this layer remains attached to the cell wall, it is usually referred to as a capsule. Many of the polymers that are secreted are, however, quite water soluble and the deposited material becomes partially detached from the cells. The more dispersed material is usually referred to as a slime layer.

role of capsules

It appears that slime layers or capsules are not essential to the cells which produce them since capsule/slime producing cells can be mutated (altered genetically) so that they no longer produce capsules or slime without any apparent harmful effects. Nevertheless with disease-causing bacteria, it is well established that capsules provide bacterial cells with protection against some animal defence mechanisms (eg ingestion by white blood cells). Exo-cellular slime layers are also involved in the movement of some prokaryotes (especially cyanobacteria) and may help to protect cells from dehydration.

negative straining

Capsules are usually observed using a negative staining technique. In this, the cells are left unstained, but the background is stained so that cells are seen in outline. The substance used for negative staining is an opaque material such as Indian ink or nigrosin which has little or no affinity for cells. Figure 1.4 outlines the negative staining procedure.

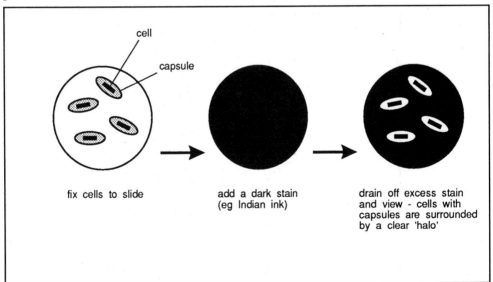

Figure 1.4 The negative staining procedure.

industrial importance of capsules and slime layers

Capsulated and slime producing prokaryotes are often responsible for the problems relating to the clogging of filters by slime in some industrial processes. The slime they produce may also choke pipes and ducts and may affect the quality of the final product. Not all exo-cellular slime production is undesirable. Some slime products are of considerable commercial value. For example, dextrans, produced as an exo-cellular slime by *Leuconostoc spp.* have been used as plasma expanders and xanthan, a polysaccharide slime secreted by *Xanthomonas spp.* has a wide variety of applications including use in cosmetics and in controlling leaks from undersea oil wells!

homo-polysaccharides

hetero-polysaccharides

The composition of capsules and slime layers varies widely. Most are polysaccharides; a few are homopolysaccharides (ie composed of only one type of monosaccharide); many are heteropolysaccharides (ie composed of more than one type of monosaccharide). Details of the chemical composition and the linkages between the monomers of these polysaccharides are known for only a few capsules and slime layers.

D-glutamic acid

The capsules of some *Bacillus spp* are comprised of polypeptides. These are, however, very unusual polypeptides. They only contain the amino acid, glutamic acid. (Compare this with the twenty amino acids found in cellular polypeptides). Also most of the glutamyl residues in these capsules are in the D (not the normal L) configuration.

Table 1.2 lists some organisms and the composition of their exo-cellular products.

∏ Use this table to assign each as either producing homopolysaccharides or heteropolysaccharides.

Organism	Name of Polymer	Composition of Exo-cellular polymer	Heteropoly-saccharide	Homopoly-saccharide
Group A				
Leuconostoc spp	Dextran	(Glucose 1-6 Glucose)$_n$		
Streptococcus spp				
Pseudomonas spp	Levan	(Fructose 2-6 Fructose)$_n$		
Group B				
Acetobacterium	Cellulose	(Glucose 1-4 Glucose)$_n$		
Agrobacterium	Glucan	(Glucose 1-2 Glucose)$_n$		
Azotobacter	Polyuronide	(Mannuronic acid-Glucuronic acid)$_n$		
Streptococcus	(Type 3)	(3 Glucuronic acid 1-3 Glucose)$_n$		
Group C				
Bacillus sp		(Glutamic acid - Glutamic acid)$_n$		

Table 1.2 Exo-cellular polymers of some bacteria.

All of the organisms, except *Azotobacter, Streptococcus* (Type 3) and *Bacillus sp,* listed in Table 1.2 are homopolysaccharide producers. This is because the exo-cellular polymers they produce contain only one type of monosaccharide (in the examples given this is either glucose or fructose).

The *Bacillus* species is, of course, not producing an exo-cellular polysaccharide it is producing exo-cellular polypeptide. The exo-cellular polymers of the *Azotobacter* and the *Streptococcus* listed are both heteropolysaccharides. You will notice that the organisms producing exo-cellular polysaccharides have been divided into two groups (Groups A and B). Those listed under group A produce the exo-cellular polysaccharides directly from exogenous substrates (eg sucrose). The synthesis of these polysaccharides is represented diagrammatically in Figure 1.5.

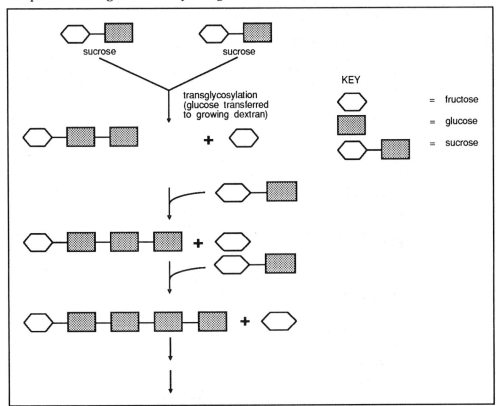

Figure 1.5 Synthesis of capsular polysaccharide from exogenous sucrose by bacteria (stylised).

SAQ 1.4

1) Using the schematic representation of dextran synthesis from sucrose given in Figure 1.5 and the structure of levan given in Table 1.2 draw a schematic representation of levan synthesis from sucrose.

2) In the text, dextran and levan were described as homopolysaccharides. Why is this not strictly true?

We have learnt that the exo-polysaccharides listed under Group A in Table 1.2 are made directly from sucrose from the medium in which the cells are suspended. Thus dextran and levan forming bacteria only produce capsules/slime layers when they grow in the presence of sucrose.

In contrast, the exo-cellular polysaccharides listed in group B in Table 1.2 are synthesised from sugars generated inside the cells. In these cases, the production of capsules and slime layers are not quite so dependent upon exogenous sugars, although we would anticipate that an abundance of sugars in the growth medium may enhance capsule/slime production.

1.4.2 Flagella

flagella

organelles

staining

Flagella (singular, flagellum) are long, thin structures that protrude through the cell wall. They are organelles (sub-cellular structures of defined function) of motility. Each flagellum is usually several times longer than the cell, but they are only about 20 nm (0.02 μm) thick. They cannot, therefore, be seen directly with a light microscope (the limit of resolution of a light microscope is about 0.2 μm). We can however, use special staining procedures to see them. This procedure often uses a dye (eg Basic Fuchsin) and Tannic acid as a fixative. The fixative causes the dye to stick to the surface of the flagellum. The flagellum, encrusted with dye, is much thicker than unreacted flagella and may just be sufficiently thick to be seen through a light microscope. Flagella are usually observed, after appropriate staining, using the higher magnification achievable in an electron microscope.

hanging drop

Indirect evidence for the presence of flagella can be obtained using a 'hanging drop' technique. This technique is outlined in the flow diagram (Figure 1.6).

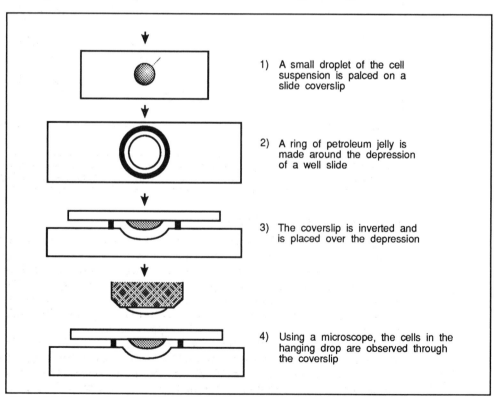

1) A small droplet of the cell suspension is palced on a slide coverslip

2) A ring of petroleum jelly is made around the depression of a well slide

3) The coverslip is inverted and is placed over the depression

4) Using a microscope, the cells in the hanging drop are observed through the coverslip

Figure 1.6 The hanging drop technique for observing motility.

runs

twiddles
Organisms with flagella show distinctive, regular directed movement interspersed with brief periods of vibrating on the spot. These two phases of movement are known as runs and twiddles. Non-flagellated organisms will be seen to show small vibrations (Brownian motion) and to remain more or less in the same place. It should be noted that not all bacteria produce flagella and those that do may only possess these structures for part of their life cycle. Flagella are quite uncommon amongst cocci, the flagella bearing bacteria are mainly bacillal (rods) and spirillal (spiral) forms.

The numbers and arrangement of the points of attachment of flagella around each cell are usually characteristic of the bacteria bearing them. The common arrangements of flagella are shown in Figure 1.7.

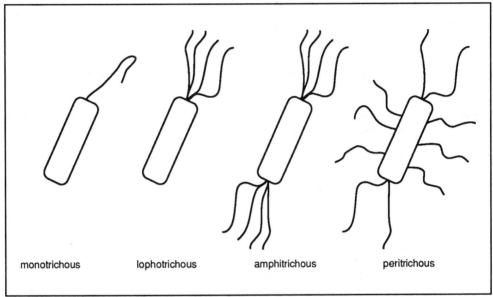

monotrichous   lophotrichous   amphitrichous   peritrichous

Figure 1.7 Arrangements of flagella around bacterial cells. Monotrichous - a single flagellum at one end of the cell, lophotrichous - several flagella at one end of the cell, amphitrichous - several flagella in two tufts, one tuft at each end of the cell, peritrichous - flagella all around the periphery of the cell.

Molecular structure of flagella

Flagella may be readily detached from cells by using the mechanical stress forces created in a blender. The detached flagella may be subsequently isolated and purified and characterised using conventional cell fractionation and analytical biochemical procedures. Flagella are made up of proteins known as flagellins. These protein sub-units of flagella are readily dissociated from each other by treatment with acid or by gentle heat treatment. Flagellins have a molecular weight of about 40,000 daltons and they are rich in the acidic amino acids, glutamic and aspartic acids. In flagella, the flagellin molecules are arranged into a helical array (see Figure 1.8) to form a long filament. These filaments are hollow and their exact form differs slightly from bacterium to bacterium both in diameter (10-20 nm) and in the 'pitch' of the helix. A few bacteria also produce flagella that are thickened by an extension of the plasma membrane (so called sheathed flagella). Some bacteria produce both polar sheathed flagella and unsheathed peritrichous flagella, a condition referred to as mixed flagellation.

flagellins

sheathed
flagella

mixed
flagellation

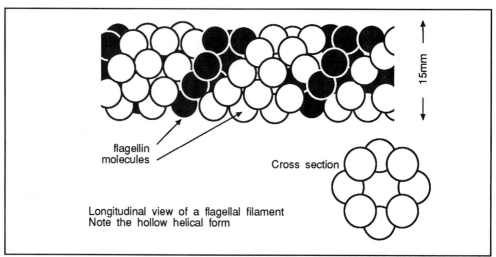

Figure 1.8 Flagellin and flagellal filaments.

basal body

The attachment of the flagellum to the cell is via a complex basal body. Only a few basal bodies have been studied in detail. Those produced by *Escherichia coli* have received most attention and their structure is represented in Figure 1.9.

The basal body consists of a central rod, inserted into a series of ring-like structures. This rod and ring structure is joined to the filament region by a slightly wider hook. This structure is typical of the flagellum of a Gram negative organism. The inner pairs of rings (S and M) are located close to the plasma membrane, the L and P rings are close to the outer layer of the cell wall.

In Gram positive cells, the cell walls are thicker and are more or less homogenous. With these cell types the flagellal basal body has only the S and M rings.

Growth of the filaments

molecular self
assembly

It is thought that flagellin is produced in the cells and that flagellin molecules are passed through the hollow core of the flagellum and are added onto its tip. (Contrast this with human hair, which grows from the base). It appears that the production of a filament from flagellin molecules occurs by a process of molecular self assembly. This means that all of the information needed to make the final structure of the filament is present in the flagellin molecules themselves. Thus once a flagellin molecule reaches the tip of the filament, it joins the filament end and automatically packs into the correct configuration. If the tip of a flagellum is broken off, it can be readily regenerated. It should be noted that some eukaryotic cells produce flagella, but the flagella are quite different from those found in prokaryotic cells.

Mechanism of flagellal movement

For many years it was thought that each flagellum moved with an undulating, wave-like movement rather like a whip. It is now widely accepted that this is incorrect and that flagella are quite rigid structures which move by rotation rather like a propeller. The rod rotates within the rings. Some flagella spin clockwise, others counter-clockwise providing propeller-like propulsion.

flagellal rotation

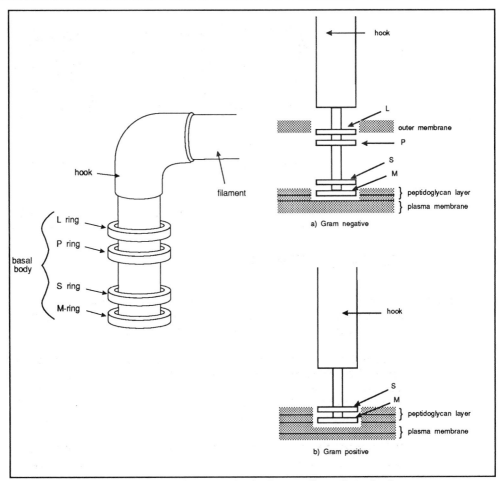

Figure 1.9 Flagellal basal bodies. a) Gram negative, b) Gram positive cells (stylised).

Cells with peritrichous flagella 'swim' in quite straight lines over some distance. These runs are interrupted by abrupt changes of direction, known as twiddles (or tumbles). It is believed that these changes are brought about by the reversal of the direction of flagellal rotation. Cells with polar flagella tend to move much faster than cells with peritrichous flagella and their motion tends to be done in many short, sharp jerks, the cells often spinning round.

1.4.3 Chemotaxis in bacteria

Whilst dealing with flagella, it is worthwhile examining the process of chemotaxis. The cells of a suspension of flagellated bacteria are usually in a state of continuous but rather random movement. Many bacteria however respond to gradients of chemicals. They may either move towards (positive chemotaxis) or away (negative chemotaxis) from the chemical. Those chemicals towards which the cells move (usually nutrients) are said to be attractants. Those chemicals they move away from (usually toxins) are referred to as repellents. Chemotaxis is readily demonstrated by immersing a small capillary tube containing an attractant or repellant into a suspension of mobile bacteria as shown in Figure 1.10.

attractants

repellants

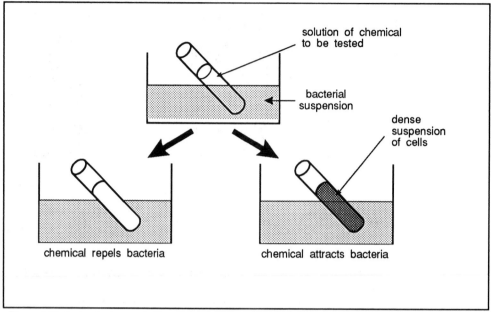

Figure 1.10 A simple demonstration of chemotaxis.

On introducing the tube into the cell suspension, the chemical in the tube diffuses out and sets up a localised gradient which either attracts or repels the cells.

In addition to being sensitive to the concentration of nutrients and toxins, some cells respond to oxygen gradients.

1.4.4 Phototaxis

It should be noted that some prokaryotic cells, particularly the photosynthetic bacteria, respond to a gradient of light intensity (ie they are said to be positively phototactic).

In a classical experiment, described in 1919 by Buder, photosynthetic purple bacteria were exposed to illumination in a spectrum. The cells accumulated at the wavelengths shown in Figure 1.11.

∏ Why do you think cells accumulated at these wavelengths and what do you think is its biological significance?

The wavelengths at which the cells accumulate are close to the absorption maxima of the main pigments (chlorophyll and carotenoids) used for photosynthesis. Biologically, it makes sense for cells which use light as a source of energy to position themselves in light at the wavelengths they can use most effectively (ie close to the wavelengths of absorption of the pigments used to absorb the light).

1.4.5 Pili

pilin

Pili have many similarities to flagella, but they are not involved in motility. They are much shorter than flagella and they are usually only about 7 nm in diameter and cannot be seen in a light microscope. Chemically they are composed of a protein called pilin.

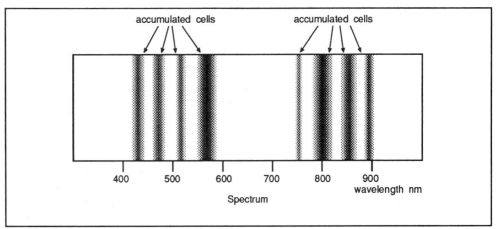

Figure 1.11 The experiment of Buder - showing the accumulation of cells in a light spectrum.

Each pilus (singular of pili) is composed of many pilin molecules associated into a straight chain (Figure 1.12).

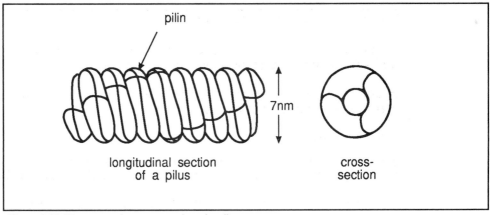

Figure 1.12 Diagrammatic representation of a pilus.

Pili are confined to certain types of Gram negative bacteria. They are thought to be mainly involved in attaching cells to substrata. Some pili (eg F pilus or sex pilus) serve as the entry point of genetic material during 'mating' amongst some bacteria. Cells which produce pili may produce many (several hundred) or as few as one.

fimbriae You should be alerted to the fact that in much of the literature, the term pili (pilus) has been used interchangeably with the term fimbriae (fimbria). Although various attempts have been made to unravel this confusion by limiting the term pilus to structures that are longer than fimbriae and are present in fewer numbers or are involved in mating, this distinction has not been universally adopted. Most authors refer to all such fine structures as pili.

We have now completed our examination of the most common structure we might find attached to, or associated with, the outside of prokaryotic cells. We can now turn our attention to the cell wall.

1.5 Cell walls

We have already examined the Gram stain. This differential stain is of great practical value for it enables us to distinguish between Gram positive and Gram negative cells (ie those that retain the primary stain of the Gram staining procedure and those that do not). In performing the Gram stain, it is important to use young, growing cells since some Gram positive cells lose their ability to retain the primary stain (ie they become

Gram variable

Gram variable). The Gram stain is thus partially conditioned by the physiological state of the cells. More importantly, it correlates well with the structure and composition of the cell walls of the eubacteria. Thus amongst the eubacteria, we find two fundamentally different cell wall types. Both of these are quite different from the cell walls of eukaryotic cells.

rigid cell walls

The cell walls of both Gram positive and Gram negative eubacteria are rigid structures which provide the cell with its shape (mycoplasma are exceptional in not producing rigid cell walls). The cell walls of prokaryotes make up 10-40% of the total dry weight of the cell and their thickness usually falls within the range 10-40 nm. Some bacteria do, however, produce considerably thicker walls.

Electron micrographs of thin sections of cell walls reveal that those of Gram positive cells consist of a single layer of almost uniform appearance. The cell walls of Gram negative cells have two readily identifiable layers (Figure 1.13).

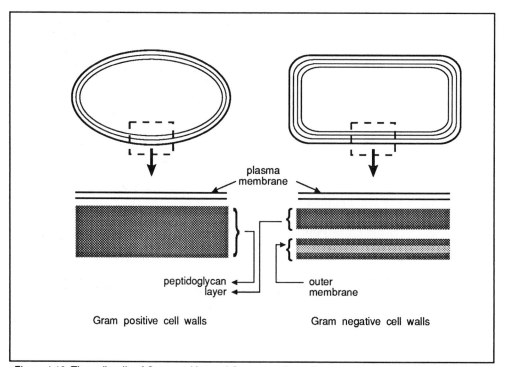

Figure 1.13 The cell walls of Gram positive and Gram negative cells.

peptidoglycan

lipopoly-
saccharides

40-90% of the dry weight of the homogenous Gram positive cell wall is made up of a complex polymer called peptidoglycan. In Gram negative cells, the peptidoglycan is confined to the inner layer of the cell wall and is present in much lower proportions. The outer layer (sometimes called the outer membrane) of Gram negative cell walls is composed of lipopolysaccharides and proteins.

1.5.1 The chemical composition of prokaryotic cell walls

murein

The rigidity of bacterial cell walls is provided by the peptidoglycan. Peptidoglycans (sometimes referred to as murein) are large polymers composed of two types of sugar derivatives, (N-acetylglucosamine and N-acetylmuramic acid), together with a small number (4 or 5) of amino acids. The most common amino acids found in peptidoglycans are L-alanine, D-alanine, D-glutamic acids, lysine or diaminopimelic acid (dap). The structure of the repeating unit of the most common peptidoglycan is given in Figure 1.14. Note that the two sugar derivatives are joined by β1-4 linkages to form a disaccharide. About 10-80 such disaccharides are coupled together to form a long strand. Such long strands are held together by the short chains of amino acids, the carboxylic acid of one peptide chain linking with the amino group of another peptide chain (see Figure 1.15). In this way a network of glycan strands inter-connected by peptide bridges is produced (see Figure 1.16).

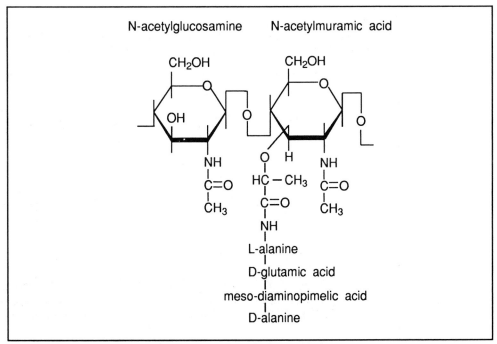

Figure 1.14 Generalised structure of the repeating unit of a peptidoglycan.

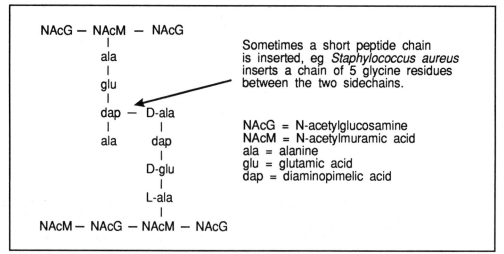

Figure 1.15 Schematic representation of the cross linkage between glycan strands through a peptide bridge. Note lysine is also a common amno acid found in the cross linkages.

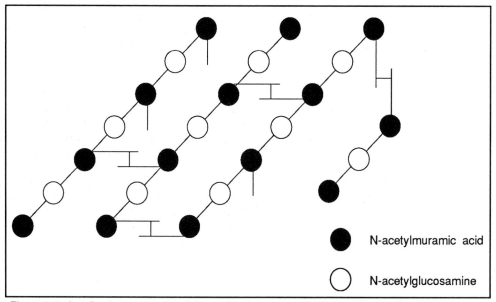

Figure 1.16 A stylised representation of the peptidoglycan network of bacterial cell walls.

The structures drawn in Figures 1.14, 1.15 and 1.16 are the most common peptidoglycan. Other arrangements are, however, known. The most common variant is the inclusion of bridging units between the peptide chains. These bridging units are usually composed of amino acids particularly glycine, threonine, serine and aspartic acid (NB aromatic amino acids, sulphur containing amino acids, some basic amino acids and proline are never found in the interstrand bridges).

N-acetyl-
talosaminuronic
acid

The 'murein' type of peptidoglycan described above is almost universal amongst prokaryotes. The principle exceptions are the archaebacteria. In these organisms, N-acetylmuramic acid is never present. It is replaced by N-acetyltalosaminuronic acid (see Figure 1.17).

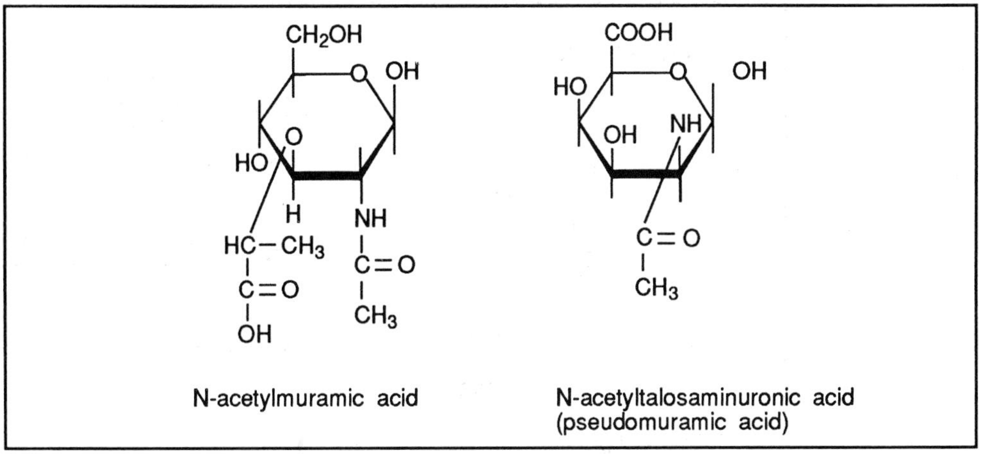

Figure 1.17 The structure of N-acetylmuramic acid and N-acetyltalosaminuronic acid.

Unlike the eubacteria, whose cell walls contain amino acids in both the D and L configuration, the archaebacteria only contain amino acids in the L configuration. The main points to remember are that peptidoglycans make up about 5-10% of the dry weight of the cell walls of Gram negative cells, but as much as 40%-90% of the cell walls of Gram positive cells. In Gram negative cells, only the inner part of the cell wall is composed of peptidoglycan. The peptidoglycans give the rigidity to both Gram positive and Gram negative cell walls.

1.5.2 The outer membrane of Gram negative cell walls

Outside of the thin murein sac of Gram negative cell walls is a structure rather similar to a typical cellular membrane. Like cell membranes, it is composed of a lipid bilayer containing phospholipids and protein. It also contains a large amount of a unique lipopolysaccharide (LPS) which probably replaces phospholipid in the outer layer.

We will be discussing the structure of membranes in detail in a later chapter, here we will give a brief overview of membrane structure. Phospholipids can be regarded as having a polar end (provided by the charge on the phosphate group) and a hydrophobic end (provided by the acyl groups of fatty acids).

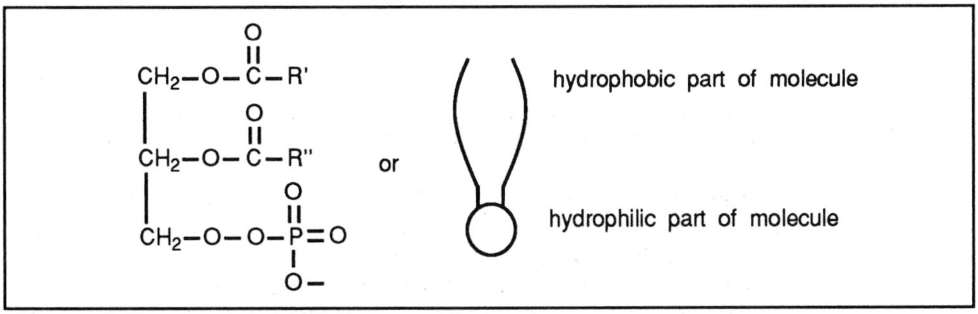

Figure 1.18 Structure of phospholipid.

When several molecules of such compounds are placed in an aqueous environment, they associate to form a bimolecular layer as shown in Figure 1.19.

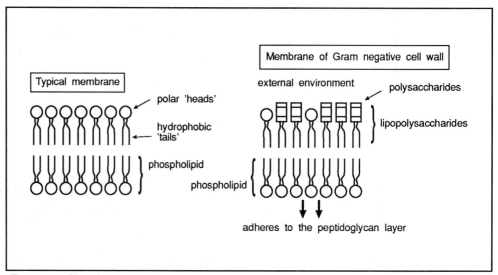

Figure 1.19 Biomolecular layers of a typical membrane and the outer membrane of a Gram negative cell wall.

In the cell walls of Gram negative cells, the phospholipids of the outer layer are replaced, at least in part, by the lipopolysaccharides.

The lipopolysaccharides of Gram negative cell walls are complex and have molecular weights of over 1000 daltons. These lipopolysaccharides are quite different even in closely related organisms. Those produced by *Salmonella spp* have been studied most thoroughly.

serotyping

[NB In members of the genus *Salmonella*, these surface lipopolysaccharides are one of the groups of compounds which make up the somatic ('O') antigens of these organisms. Identification of *Salmonella* strain, usually involves using antibodies to identify these antigens. Over 1300 different serotypes of *Salmonella* are known and, although these difference are not all attributable to the surface lipopolysaccharides, this large number indicates some wide variations in lipopolysaccharide structure].

A typical Gram negative cell wall lipopolysaccharide is provided in Figure 1.20. Do not attempt to remember all of the details of this structure but do note that the structure can be divided into three regions, Lipid A, an R core region and an O side chain.

porins

specific
channel
proteins

A number of proteins are also associated with the outer membrane of Gram negative cell walls. Many of these are involved with the transport of materials towards the cells. Some produce small pores in the outer membrane. These are called porins. Others specifically bind chemicals and transport them through the cell wall (specific channel proteins); LamB for example enhances the diffusion rates across the walls of *Escherichia coli* of maltose and maltodextrin.

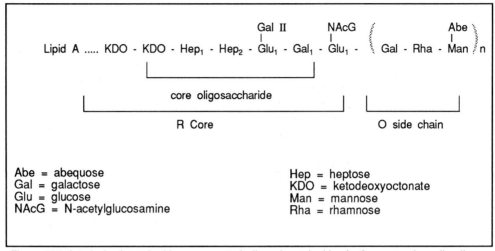

Figure 1.20 A typical polysaccharide component of a lipopolysaccharide of a Gram negative cell wall.

The outer membrane of Gram negative cells is tied to the inner peptidoglycan layer through a number of protein cross links between the peptidoglycan and the outer membrane. The overall structure of a cell wall of a Gram negative cell is represented in Figure 1.21.

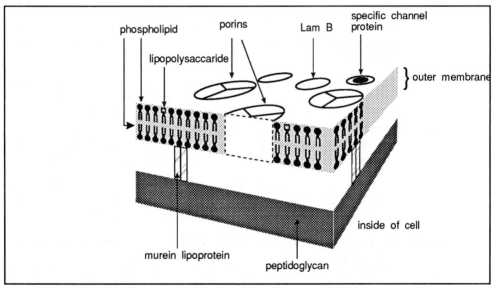

Figure 1.21 An overall representation of the cell walls of a Gram negative cell.

1.5.3 The cell walls of Gram positive bacteria

We have learnt that the cell walls of Gram positive organisms appear as a uniform layer around the cell and that the major component of this layer is peptidoglycan. As much as 90% of the dry weight of the cell wall can be made up of this single component. Gram positive bacteria cell walls do, however, contain some other components including teichoic acid   polysaccharides and a group of polymers known as teichoic acids. The structure of some teichoic acids are included in Figure 1.22. If you examine these structures carefully you will see that many of them contain glycerol phosphate or ribitol phosphate as a core component. D-alanine is also a common component. The majority of the teichoic acids appear to be exposed on the outer surface of the cell wall.

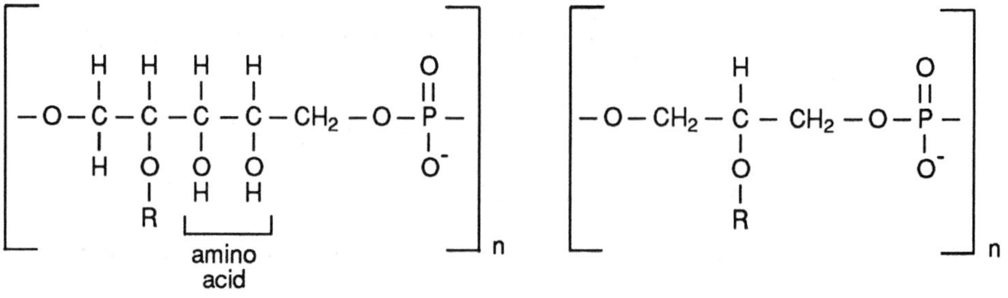

ribitol based teichoic acid

glycerol based teichoic acid

R = various residues depending on source - glucose and succinate are quite common.

Figure 1.22 Core structure of some teichoic acids.

Π In attempting to learn the basic facts about the cell walls of various groups of prokaryotes, you might find it useful to construct your own kind of revision table, perhaps using the following format:

Characteristic	Gram positive	Gram negative	Archaebacteria
Thickness			
Homogenous			
N-acetylmuramic acid present			
amino acids present			

When you have done this, attempt the next SAQ.

1.5.4 Protoplasts and spheroplasts

The enzyme lysozyme is known to hydrolyse the peptidoglycans present in bacterial cell walls. In other words, this enzyme breaks up the long strands in the peptidoglycan layers of the cell wall.

SAQ 1.5	Assign each of the following characteristics to the cell walls of Gram positive, Gram negative bacteria or archaebacteria.

1) Under an electron microscope, the cell wall appears as a more or less homogenous layer.

2) Under an electron microscope, the cell wall appears to have an outer membrane layer.

3) The cell walls contain up to 90% peptidoglycan by dry weight.

4) The cell walls contain N-acetylmuramic acid.

5) The cell walls contain N-acetyltalosaminuronic acid.

6) The cell walls contain some D amino acids.

7) Ribitol teichoic acids are present.

8) Substantial amounts of phospholipids and lipopolysaccharides are present.

9) In a basal body of a flagellum inserted in the cell wall, four sets of rings can be seen under an electron microscope.

10) The cell walls contain N-acetylglucosamine.

∏ What do you think would be the consequences to bacterial cells if they are exposed to this enzyme?

Since the rigidity of bacterial cell walls can be attributed to the peptidoglycan, breakdown of the peptidoglycan around the cell would cause a loss of this rigidity. The tendency is for the cells to quickly lose their shape and become spherical. Continued treatment with lysozyme causes the peptidoglycan to become completely hydrolysed. The smaller fragments thus formed just float away from the cell and the cell becomes stripped of its cell wall to leave a naked protoplast. Such protoplasts are very fragile since they are only surrounded by a delicate plasma membrane. If these protoplasts are suspended in dilute salt or sugar solutions, they quickly take up water by osmosis. The protoplast lysis   protoplasts swell and burst (lyse). If, however, the cell walls are removed while the cells are suspended in a solution of a suitable chemical (sucrose is commonly used) which has a similar concentration to the solutes inside of the cell, the protoplasts that are formed are stabilised.

∏ What would you anticipate would happen if such protoplasts were transferred 1) into distilled water or 2) into a very concentrated solution of sucrose?

We should perhaps expect that protoplasts which are transferred to distilled water would quickly undergo lysis (bursting) because they would take up water by osmosis and thus swell stretching the plasma membrane. If, on the other hand, we transferred the protoplasts into a very concentrated solution of sucrose, water would move, by osmosis, from the protoplast into the external solution. Under these conditions the plasmolysis   protoplasts would shrivel or collapse - a process called plasmolysis.

We can summarise what we have in learnt about the preparation of protoplasts in Figure 1.23.

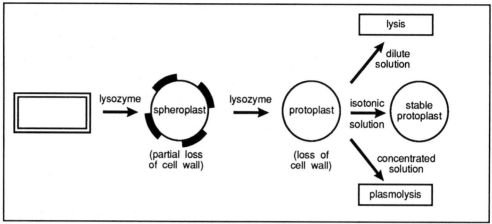

Figure 1.23 Protoplast preparation and stabilisation.

Protoplasts that are kept in appropriate salt concentration are always spherical, irrespective of the original shape of the cells from which they are derived. Strictly speaking, protoplasts are structures which have *all* of the cell wall components removed. However if the cell wall is only partially removed, but its rigidity is lost, then a spherical structure will be produced. Such structures are called spheroplasts. The most common reagent used to produce protoplasts is the enzyme lysozyme. This enzyme, produced widely in nature (it is found for example in tears and in egg whites), hydrolyses the glycosidic links between N-acetylmuramyl and N-acetylglucosamine in the core peptidoglycan of the cell wall. Protoplasts and spheroplasts have proven to be useful for studying processes in which cell walls may interfere (eg naked cells may take up exogenous DNA more readily than native cells). They have also been useful in the study of how cell walls are produced.

spheroplasts

1.5.5 Cell wall synthesis, L-forms and mycoplasma

For the most part we would anticipate that bacteria could not survive in nature without their cell walls. We would also anticipate that as cells grow and divide, there must be cell wall synthesis. What we might not expect is that for a cell to synthesise peptidoglycan, peptidoglycan must already be present.

In other words, for a cell to join N-acetylmuramyl and N-acetylglucosamine residues to produce peptidoglycan strands and to insert these into cell walls, they can only do so if peptidoglycan molecules are already present in the wall. Thus if no peptidoglycan molecules exist on the surface of a cell, no further peptidoglycan can be produced. This is important because it has some quite significant consequences.

Mycoplasma are essentially free-living 'protoplasts'. The reason they can survive without a cell wall is because they live in environments with an osmotic pressure similar to that of protoplasm (eg within animal bodies) and because they have developed rather special plasma membranes.

Also interesting in this context, are the L-forms of bacteria. First discovered in 1935, L-forms appear to be bacteria which no longer produce cell walls. They are called L-forms because they are much larger (ie Large forms) than the normal wall-producing cells from which they are derived. L-forms may arise spontaneously by continuously cultivating cells on rich medium (this was how L-forms were first discovered, with the bacterium *Streptobacillus moniliformis*). More frequently, L-forms can be selected for by cultivating bacteria in an osmotically buffered medium containing penicillin. Penicillin is known to inhibit cell wall (peptidoglycan) synthesis. Usually, if the penicillin is removed soon after L-forms are first formed, then the cells rapidly revert to their normal shape. If, however, the L-forms continue to be cultivated for a long time in the presence of penicillin, the cells may continue to grow in the L-form even when they are transfered to penicillin-free medium (ie the L-forms are stable).

∏ Can you explain why L-forms, produced by growing cells in the presence of penicillin for a long period, may be stabilised?

primer for peptidoglycan synthesis

The most likely explanation is that short term incubation in penicillin, by inhibiting peptidoglycan synthesis, leads to a reduction in the amount of peptidoglycan in the cell wall and hence L-forms are produced. When the inhibition of peptidoglycan synthesis by penicillin is removed, there is still sufficient peptidoglycan present in the cell wall to act as a primer for peptidoglycan production. Under these conditions normal cell walls can be re-established and the cells return to their normal shape and size.

If however incubation in penicillin is prolonged, the amount of peptidoglycan around each cell wall becomes progressively less and less until it becomes insignificant. Once this stage is reached, the cells can never synthesize cell wall peptidoglycan because there are no 'primer molecules' left in the cell wall. Thus the L-forms breed true whether or not penicillin is included in the medium.

1.6 The plasma membrane

We will discuss membranes in detail in Chapter 3, but for completeness we include a brief description of prokaryotic plasma membranes here. The plasma membrane, also called the cell membrane or cytoplasmic membrane or plasmalemma, lies just beneath the cell wall. It is a flexible structure and, based upon electron micrographs, is about 7.5 nm thick. This is a vital structure. If it is broken, the contents of the cell spill out and usually cell death ensues. The structure of the plasma membrane has similarities with the outer membrane of Gram negative cell walls in so far as it mainly consists of a molecular bilayer of phospholipids. It has therefore the typical unit membrane structure (see Chapter 3). The bimolecular layer of phospholipids is stabilised by the cations Ca^{2+} and Mg^{2+} which associate with the negative charge on the phosphate groups.

hopanoids

The plasma membranes of prokaryotes, however, differ from the membranes of eukaryotes in one important respect. In eukaryotes, sterols, especially cholesterol, are embedded in the lipid layers of the unit membrane. These plate-like molecules act as 'stiffeners' in this rather fluid-like layer. Prokaryotes contain very little sterol, instead they use hopanoids or squalene to act as membrane 'stiffeners' (Figure 1.24).

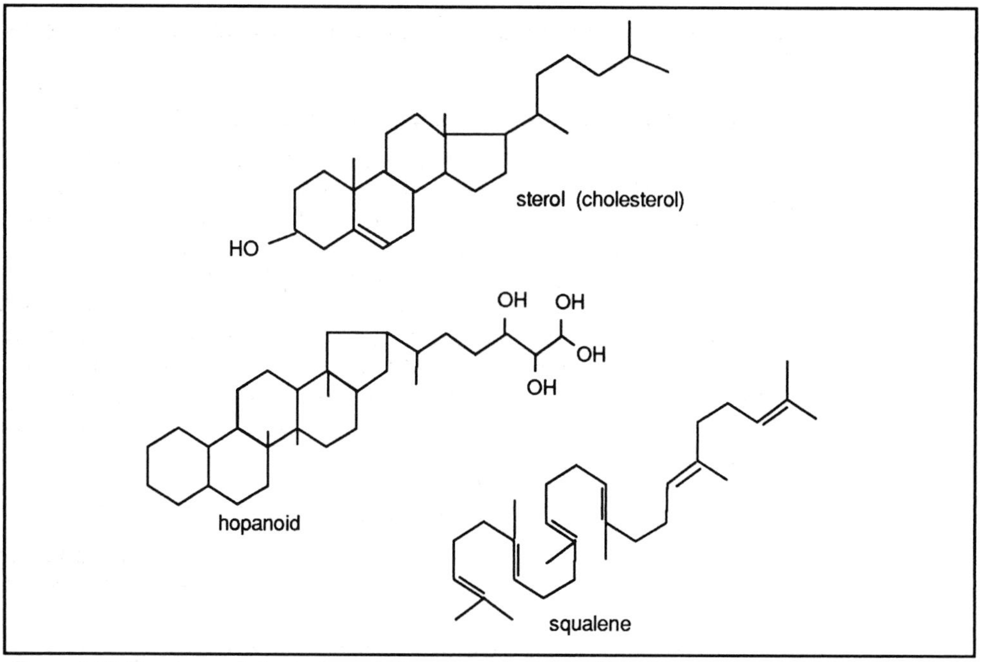

Figure 1.24 Structures of sterols, hopanoids and squalene found in unit membranes.

The membranes of archaebacteria also differ from the membranes of eubacteria and of eukaryotes in the composition of the phospholipids. The phospholipids of eubacteria and eukaryotes contain fatty acids which are almost always linear acyl fatty acids such as palmitic, oleic, stearic and linoleic acid (Figure 1.25). In archaebacteria, the phospholipids of the plasma membrane contain branched fatty acids containing a variable number of isoprenoid units (Figure 1.25).

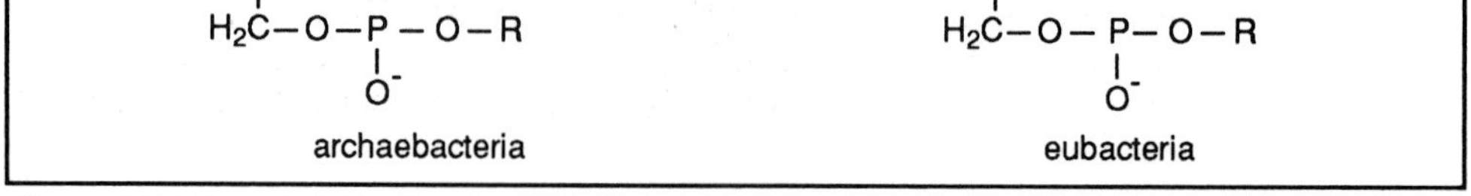

Figure 1.25 Phospholipids of archaebacteria and eubacteria.

1.6.1 The roles of the plasma membrane in prokaryotes

Although some small non-polar (ie fat soluble) substances might penetrate the membrane by dissolving into the lipid phase of the membrane, most ionised (polar) molecules will not pass the membrane barrier by simple, passive diffusion. The plasma membrane acts, therefore, as a general barrier. Passage of material into and out of the cell depends upon modifications to the structure of the membrane and the presence of

permease

specific transport proteins (permeases). More details of these are provided in Chapter 3.

The plasma membrane as a centre for generation of ATP

The main energy commodity of biological systems is adenosine triphosphate (ATP). The production of ATP in aerobic bacteria takes place in association with the plasma membrane. Thus, in those prokaryotes which carry out respiration (ie those that oxidise organic nutrients to produce cellular energy in the form of ATP), we find the agents of respiration (electron transport components; phosphotransferases) integrated into the structure of the plasma membrane. In some photosynthetic bacteria (eg purple photosynthetic bacteria), the complete apparatus for the harvesting and conversion of light energy into cellular energy (ATP) is located in the plasma membrane. We will examine the process of ATP generation during respiration and photosynthesis in more detail in Chapter 4. The reader is also referred to BIOTOL text 'Principles of Cell Enegetics'.

The plasma membrane as a centre of biosynthetic activity.

The enzymes involved with the synthesis of membrane lipids and peptidoglycan and other cell wall components are also associated with the plasma membrane.

1.6.2 Modifications of the plasma membrane

mesosomes

Generally, the plasma membranes follow the contours of the cell wall. Sometimes however the plasma membrane invaginates to form complex membranous structures called mesosomes. Mesosomes are always continuous with the cytoplasmic membrane and they are usually found at the point where the membrane starts to fold in, prior to cell division. They are thought to be involved with the division of nuclear material at the time of cell division and are the focus for DNA replication (Figure 1.26).

1.6.3 Internal membranes

If respiration and photosynthesis in prokaryotes occur in or on the plasma membrane, how can higher rates of respiration and photosynthesis be achieved? The answer is, by extending the surface area of the plasma membrane by infolding it into the cytoplasm. Organisms like *Azotobacter*, which have especially high rates of respiration, greatly enlarge the total area of their plasma membranes by having a considerable amount of infolded membrane.

thyllakoids

The membranes which intrude into prokaryotic cells are, for the most part, continuous with the plasma membrane at the periphery of the cell. It does appear, however, that the membrane carrying the photosynthetic apparatus in the cyanobacteria (formerly called the blue green algae) might be flattened, membranous sacs suspended in the cytoplasm. These sacs (called thyllakoids) have only rarely been observed to be connected with the plasma membrane.

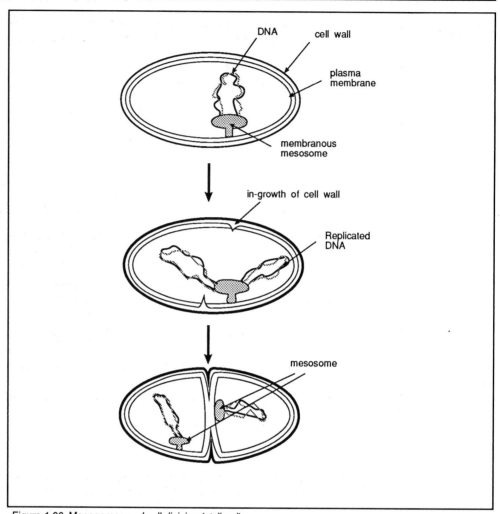

Figure 1.26 Mesosomes and cell division (stylised).

Let us now attempt to answer some questions concerning prokaryotic plasma membranes and cell walls. Attempt SAQ 6.1 before moving on.

1.7 Prokaryotic protoplasm

nuclear region
cytoplasm

Protoplasm is the name given to the whole of the cell contents bounded by the plasma membrane. Traditionally protoplasm is divided into two components - the cytoplasm and the nucleus. The nucleus is that part of the cell which carries the genetic information, while the cytoplasm is generally considered to be the chemical factory in which metabolism takes place. In the conventional sense, prokaryotes do not possess nuclei. They do however contain deoxyribonucleic acid (DNA). This is usually present as an irregular shaped body embedded in the cytoplasm. Unlike eukaryotes, this region is not separated from the cytoplasm by a membrane. Thus with prokaryotes we should not think about distinctive nuclei, but should remember that prokaryotic cells contain what we might call nuclear regions.

SAQ 1.6

1) The membranes from a population of cells have been analysed and shown to contain the following chemicals: cholesterol; phospholipids containing stearic acid residues; proteins.

Which of the following groups of organisms do these cells probably belong?

Eubacteria, mycoplasmas, cyanobacteria, eukaryotes, archaebacteria.

2) Assign the following structures to the group or groups of organisms in which they are usually found.

Structure	Groups of organisms
Mesosomes	Gram positive eubacteria
Hopanoids	Gram negative eubacteria mycoplasmas archaebacteria
Phospholipids containing isopentanyl derivatives	cyanobacteria azotobacters eukaryotes

3) What would probably happen if penicillin is added to:

a) a growing culture of eubacteria in a dilute medium;

b) a non-growing suspension of eubacteria in a dilute medium;

c) a growing culture of L-form bacteria.

The cytoplasm (ie the rest of the protoplasm excluding the nuclear region) of prokaryotes has often been described as being rather homogenous. This is not strictly true; we shall learn shortly that several different types of structures may be embedded in the cytoplasm. It is true, however, that at most magnifications, the bulk of the cytoplasm is of uniform appearance. Let us examine the prokaryotic nuclear regions and cytoplasm in a little more detail.

1.7.1 Nuclear region

To demonstrate the nuclear region of prokaryotic cells using a light microscope, basic (cationic) dyes can be used. These stain both DNA and RNA. The cytoplasm of prokaryotic cells contain many ribosomes. These particles are part of the machinery of the cells involved in protein synthesis and are composed of about 60% RNA and 40% protein. Thus staining the cells with basic dyes, results in the whole of the cytoplasm and the nuclear region becoming almost evenly stained. If however the cells are pre-treated with dilute acid or alkali the less stable RNA is hydrolysed and washed out of the cells without damage to the DNA. If such cells are now incubated with a basic dye, the dye associates with the DNA left in the cells. It is however more usual to examine the nuclear region using an electron microscope. The nuclear regions of prokaryotic cells are small and yet packed into these small areas is all of the cell's genetic information. In 1963, in what was to become a classic study, Cairns succeeded in gently breaking open bacterial cells to release intact DNA molecules and to spread them out. From this study (and later confirmed by many other studies) it was shown that the DNA

DNA packaging

was in the form of a circular thread. For *Escherichia coli* the thread was estimated to be about 1200 μm long (ie about 600 times longer than the cell which contained it!) Clearly in the cell, the DNA is much twisted up on itself. It is thought that this folding may be helped by neutralisation of the negative charges on the phosphate groups of the DNA by Mg^{2+} ions, basic (positively charged) proteins and polyamines. RNA has also been implicated in the folding and packing of DNA in prokaryotic cells.

1.7.2 Prokaryotic cell cytoplasm

The principle pre-occupation of growing cells is with making proteins. In a typical culture of bacteria, about 80% of the total metabolic activity of the cells may be directed towards this end. To make proteins, the cell has to make the appropriate amino acids and generate the energy to fuel this synthesis. In addition, the machinery (transfer RNA and ribosomes) for making proteins has also to be made available. We will discuss protein synthesis in more detail in a later chapter. Many bacteria can double their weight in less than half an hour. It is not surprising therefore to find that these cells contain enormous numbers of ribosomes and that we can think of the cytoplasms of prokaryotic cells as being intense protein factories.

70s ribosomes

The ribosomes found in prokaryotes are quite different from those found in eukaryotes. Firstly, they are smaller. They sediment in an ultracentrifuge with a sedimentation coefficient of 70s (s = Svedberg unit) compared with those of eukaryotes (sedimentation coefficient of 80s). They also have quite different sensitivities to chemicals (ie they are functionally different). Many of the antibiotics used to treat bacterial infections inhibit protein synthesis by 70s ribosomes but not by 80s ribosomes.

The cytoplasm of prokaryotic cells also contain many of the enzymes involved in catabolism (breakdown of nutrients) and in the synthesis of substances needed for growth (eg amino acids, sugars, lipids, nucleotides).

1.8 Cytoplasmic inclusions

Although the cytoplasm of prokaryotic cells is generally considered to be a rather uniform matrix, it may also include other structures. The largest of these may be visible at maximum magnification in a light microscope. Usually, however, they only become visible using the greater magnification achieved using the electron microscope.

There are many different cell inclusions. Here we confine ourselves to:

*   storage materials;
*   gas vesicles and vacuoles;
*   magnetosomes;
*   chlorosomes;
*   polyhedral bodies (carboxysomes);
*   endospores.

1.8.1 Storage compounds

poly-β-hydroxybutyrate

Poly-β-hydroxybutyrate is the most common prokaryotic storage compound which is deposited in sufficient quantities to be seen under a light microscope. It is a polymer of hydroxybutyric acid joined by ester links (Figure 1.27). These polymeric molecules aggregate to form granules.

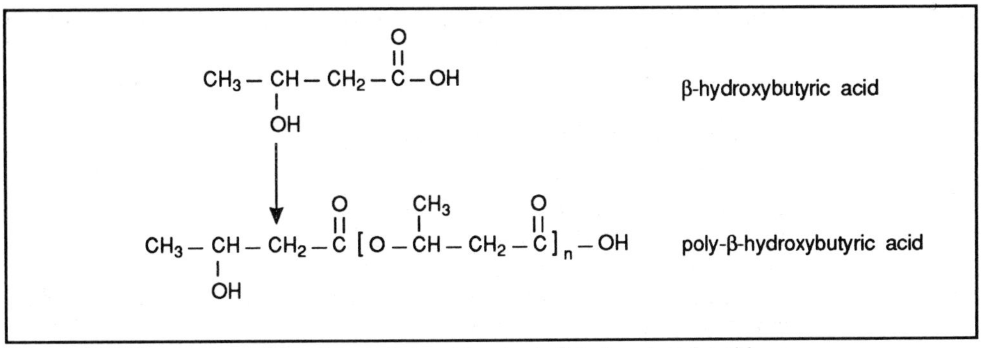

Figure 1.27 The structure of β hydroxybutyric acid and poly-β-hydroxybutyric acid.

These granules stain well with fat soluble dyes such as Sudan Black. Poly-β-hydroxybutyrate is readily metabolised by the host cell during times of nutrient starvation to provide both cellular energy and carbon. Attempts have been made to use bacterially produced poly-β-hydroxybutyrate as the basis for the production of biodegradable 'plastics'.

glucans   Glucans are polymers of glucose which some prokaryotes use as a storage compound in much the same way that animals store glycogen and plants store starch. Bacterial glucan deposits are usually smaller than poly-β-hydroxybutyrate granules.

polyphosphate   Polyphosphate granules are produced by many microbial cells when the cells are grown in phosphate-rich environments. These polymeric phosphate compounds are readily

metachromatic granules   stained with basic dyes. Polyphosphate granules are sometimes known as metachromatic granules because they exhibit colour changes (metachromasia) appearing red/violet when stained with the dye Toluidine Blue.

cyanophycin   Cyanophycin is a co-polymer of the amino acids arginine and aspartic acid which acts as a reserve of nitrogen. It is produced by some cyanobacteria in which it can make up as much as 10% of the dry weight of the cell. In general, however, prokaryotes do not store intracellular nitrogen reserves.

1.8.2 Gas vesicles and vacuoles

Most cells are slightly denser than water and hence photosynthetic cells growing in an aqueous environment need a flotation device in order to remain close to incoming light at the water surface. These cells often trap air in gas vacuoles within the cells as boyancy devices. Under the electron microscope, the gas vacuoles are seen to be made up of a number of small gas vesicles (Figure 1.28). Each vesicle is generally shaped as a long hollow cylinder, about 7.5 nm in diameter and about 200-1000 nm long. The borders of the vesicles are composed of tightly packed proteins. It is believed that water is excluded from the interior of the vesicles during their formation. It can be shown by rapidly changing the hydrostatic pressure of the cell suspension that collapsed vesicles never recover. Cells can only re-acquire gas-filled vesicles by the production of new vesicles.

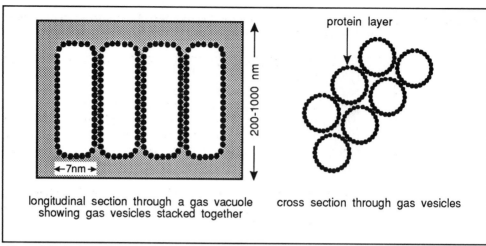

Figure 1.28 Prokaryotic gas vesicles.

1.8.3 Other cytoplasmic inclusions

Magnetosomes

magnetosomes

In 1975, R P Blackmore described the magnetic behaviour of some bacteria. These, cells when placed in quite weak magnetic fields, move to one or the other magnetic poles. The cells contain small enclosed crystals of magnetite (Fe_3O_4) which act as the sensing agents. These magnetosomes, as they are called, seem to be capable of orientating cells in a magnetic field.

Magnetotactic organisms have been found in many aquatic environments. Interestingly those from the northern hemisphere almost all migrate towards the north pole while those isolated from the southern hemisphere seek the south pole. It is thought this behaviour helps organisms that have become accidentally suspended in a column of water to re-find the bottom where the physical environment and available nutrients are more suitable for survival. It is thought to work in the following way. The Earths' magnetic field forms an angle with the surface. In the northern hemisphere, the field is inclined downwards, in the southern hemisphere upwards. Thus an organism suspended at any point in the northern hemisphere would simply have to swim along the magnetic lines of force to return to the bottom. Those suspended in the southern hemisphere would have to swim in the opposite direction. Such an explanation is consistent with the differences in magnetotactic behaviour observed in isolates from the northern and southern hemispheres.

Chlorosomes

chlorosomes

Chlorosomes are composed of distinctive cigar shaped vesicles which house the photosynthetic apparatus. These small structures (50 x 100 nm) are confined to green photosynthetic bacteria.

Polyhedral bodies (carboxysomes)

carboxysomes

Many cells which use carbon dioxide as a source of carbon for cell synthesis contain polyhedral bodies which house the apparatus (enzymes) involved in the assimilation of carbon dioxide. These structures are also known as carboxysomes.

| **SAQ 1.7** | 1) Which of the following list of cellular compounds would readily take up basic dyes?

Polyphosphate, DNA, RNA, poly-β-hydroxybutyrate, acid proteins, basic proteins.

2) When prokaryote A is grown on medium 1 it is shown to contain cytoplasmic granules which stain densely with Sudan Black. If on the other hand the same organism is grown on a different medium (medium 2), no such granules can be observed. What is the most likely interpretation of these observations?

3) Organism B contains metachromatic granules when grown on medium 3 but not when grown on medium 4. What is the most likely explanation of these observations?

4) A suspension of a cyanobacterium was held in a bottle fitted with a rubber bung. The rubber bung was struck with a hammer. Quite quickly the cells fell to the bottom of the bottle but after about 20 minutes they began to rise. About 1 hour after the bung had been struck, all of the cells had accumulated close to the top of the vessel. Explain why this sequence of events took place. |
|---|---|

1.9 Endospores

We complete this chapter by a discussion of endospores. In the preface we referred to the fact that prokaryotic cells generally do not undergo differentiation (ie each species only produces one cell type). This is a reflection of the fact that most are unicellular (each organism consisting of only one cell) and therefore each cell has to be capable of carrying out all of the functions necessary for the cell to survive and reproduce. Eukaryotes on the other hand are often multicellular and show a high degree of cell specialisation *differentiation* (differentiation). The divisions between unicellular and multicellular and non-differentiating and differentiating systems are not entirely confined to the distinction between prokaryotic and eukaryotic cell types. Thus many eukaryotes are also unicellular and some prokaryotes can exhibit some degree of multicellularity and of cell specialisation. The most striking, and common, condition of differentiation shown by prokaryotic cells is in the transition from actively growing and metabolising cells (vegetative cells) into metabolically inert forms (spores). We will return to the problems of multicellularity in later chapters, here we will confine ourselves to a discussion of spore development.

The production of spores is widespread but not universal amongst prokaryotes. Those that produce spores often do so under conditions unfavourable for growth. The spores *thermoduric* produced are usually extremely resilient being tolerant to heat (ie they are thermoduric), desiccation, disinfectants and radiation. When moved to favourable conditions, they germinate and produce vegetative cells. Thus spores can be considered as resting or resistant forms for the survival of the species during conditions which do not favour the growth of vegetative cells.

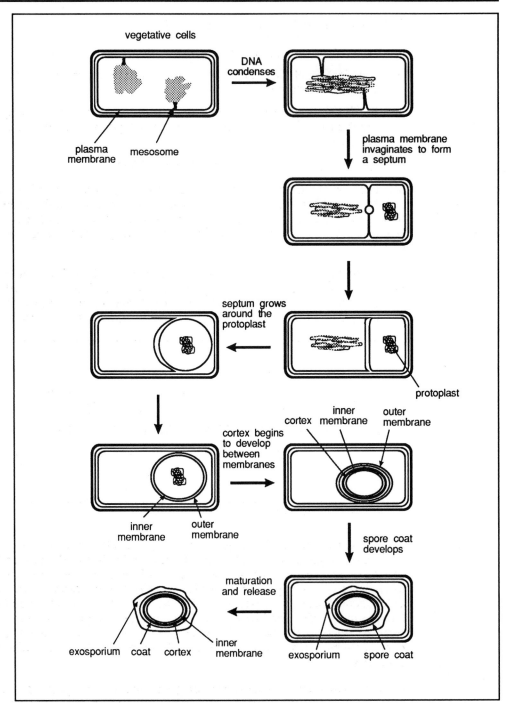

Figure 1.29 Structural changes during bacterial sporulation.

Most commonly, prokaryotes produce spores within vegetative cells and the spores are consequently usually referred to as endospore. Figure 1.29 gives a schematic representation of the structural changes which occur during sporulation.

The major steps in this sequence are:

- the condensation of the DNA to form a filament;

- invagination of the plasma membrane to form a septum (this usually occurs close to one end of the cell);

- development of several layers around the septum to form initially a spore cortex and then a multi-layered coat. The outmost layer is called the exosporium; (plurul exosporia);

- lysis of the parent cell and release of the spore.

terminal, subterminal and central spores

The position of the spore in the cell, differs amongst different cell types. Spores may develop towards one end (terminal spores), more-or-less central (central spores) or just to one side (subterminal spores). The diameter of the spore may be larger or smaller than that of the vegetative cell which produced it. The parent cell might therefore become distended.

dipicolinic acid

Under the light microscope, spores are highly refractile and take up stains much less readily than vegetative (growing) cells, reflecting their quite different chemical compositions. Endospores contain large (5-10% dry weight) quantities of a compound called dipicolinic acid (Figure 1.30). This acid associates with Ca^{2+} ions.

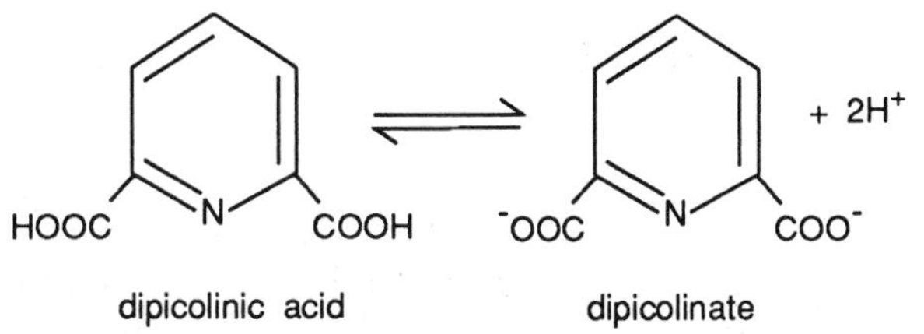

$$\text{dipicolinic acid} \rightleftharpoons \text{dipicolinate} + 2H^+$$

Figure 1.30 Structure and dissociation of dipicolinic acid.

cortical layer

Dipicolinate is not found in vegetative cells. Spores also contain substantial amounts of Ca^{2+} which, together with peptidoglycan and dipicolinic acid make up the bulk of the cortical layer. It is thought that the presence of dipicolinic acid and low water content of spores is, at least in part, responsible for the heat tolerance shown by these structures. Often sporulation is accompanied by the production of antimicrobial substances (compounds which inhibit/kill microorganisms). Examples of these are some cyclic peptides such as the bacitracins which inhibit cell wall synthesis, gramicidin-type peptides which can modify membrane structure and the edeins which inhibit DNA synthesis.

bacitracins

gramicidin

edeins

spore activation

Mature spores have the capability of germinating to produce vegetative cells. This involves an activation stage. Activation is usually by an environmental trigger such as heat (eg 60-70° for a few minutes). This is rapidly followed by the spores becoming less refractile, less heat resistant and more ready to take up dyes. The spores subsequently swell and the coat ruptures and a new cell emerges through the spore coat - a process called outgrowth.

outgrowth

Endospores of the type described above can be found amongst Gram positive organisms including strict aerobes (eg *Bacillus spp.*) and strict anaerobes (eg *Clostridium spp.*). It should be noted that some prokaryotes (Streptomyces) produce spores **exospores** externally (exospores). These organisms produce long filamentous cells (called hyphae). During the formation of exospores, cross walls divide the hyphae into individual 'cells' which round up and produce thickened walls and become dormant (Figure 1.31).

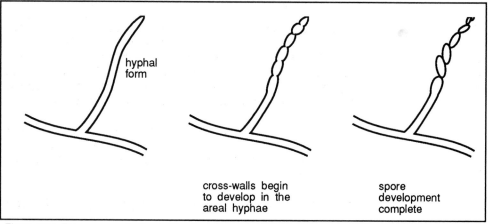

Figure 1.31 Exospore production in Streptomyces.

SAQ 1.8

Using the lists provided, assign functions to the following prokaryotic structures. (NB some structures may have more than one function).

	Structure	Functions
1)	Pilus	Energy and carbon reserves.
2)	Flagellum	Directly involved in protein synthesis
3)	Mesosome	Storage of phosphate
4)	70s ribosomes	Involved in attachment of cells to a substratum
5)	Metachromatic granules	Involved in dispersal
6)	Peptidoglycan layer	Rigidity of cell wall
7)	Polyhydroxybutyrate granuoles	Involved in DNA synthesis
8)	Endospore	Survival in hostle environments
9)	Capsules	A focus for cell division

Summary and objectives

In this chapter we have explored the structure of prokaryotic cells, beginning from the outside and working inwards. We have learnt that outside of the cell may be organelles of motility (flagella) and attachment (pili) as well as capsules and slime sheaths. We have also examined the organisation and composition of the cell walls of prokaryotes and have learnt that they can be divided into a few fundamentally different types.

We then turned our attention to the cell interior and learnt about the plasma membrane, and a variety of intra-cellular structures. We did not exclusively concentrate on their structure but also gave some attention to their function.

Now that you have completed this chapter you should, therefore, be able to:

- describe prokaryotic cells in terms of their size, shape and arrangement. From the arrangement of cells, you should also be able to describe the mode of cell division.

- explain why the small sizes of prokaryotic cells may account for their very fast rates of metabolism and growth.

- explain why many of the sub-cellular structures found in prokaryotic cells cannot be seen by light microscopy.

- identify and label the major sub-cellular structures found in prokaryotic cells.

- assign functions to the sub-cellular structures found in prokaryotic cells.

- describe the strategies adopted for staining prokaryotic cells and sub-cellular structures.

- describe the differences in Gram positive, Gram negative and archaebacterial cell walls in terms of physical appearance and chemical composition.

- describe the processes of producing protoplasts, spheroplasts and L-forms.

The organisation of eukaryotic cells

The organisation of eukaryotic cells

2.1 Introduction

Multicellular animals and plants are the most prominent complex organisms found on Earth and they all contain cells which show the eukaryotic rather than the prokaryotic type of cellular organisation. No prokaryotic organism has ever developed the complexity of multicellular form characteristic of eukaryotes. Although some produce colonies, there is very limited specialisation or division of labour. The evolution of eukaryotic organisation was a breakthrough of dramatic proportions as far as life on Earth is concerned.

What is so special about eukaryotic organisation?

In this chapter we will examine the structure of eukaryotic cells, in comparison with prokaryotic cells, to note the differences and to see if we can find clues as to why eukaryotic but not prokaryotic cells gave rise to multicellular organisms.

2.2 Animal cell structure

true nucleus

The term eukaryote means 'true nucleus' and refers to the fact that the genetic material, DNA, is contained within a distinct and discrete zone surrounded by a membrane. This is the nucleus (Figure 2.1). The nucleus is usually the most striking distinguishing feature of eukaryotic cells but it is not the only one. Examination with the electron microscope reveals the presence of a large number of membranes forming interior organelles
compartments called organelles, which occupy a large proportion of the space between the nucleus and the cell membrane.

The nucleus is the largest organelle and is approximately 5μm in diameter. This is larger than many prokaryotic (bacterial) cells. It is not surprising, therefore, to discover that eukaryotic cells are considerably larger than prokaryotic cells. The DNA in the nucleus is readily stained and this, coupled with its size, makes the nucleus easy to see using the light microscope. Several other particles can be seen with the light microscope but it is only with the electron microscope that details of their structure can be discerned. Figure 2.1 shows a representation of what can be seen in an animal cell using the electron microscope.

plasma membrane

The cell is bounded by an outer membrane, the plasma membrane (or plasmalemma) but in animal cells there is no cell wall. For this reason students sometimes refer to the plasma membrane as the 'wall' of the animal cell. This can lead to confusion when discussing plant cells because plant cells have a cell wall separate from the plasma membrane. It is recommended that the term cell wall is retained for a structure external to the limiting membrane of the cell.

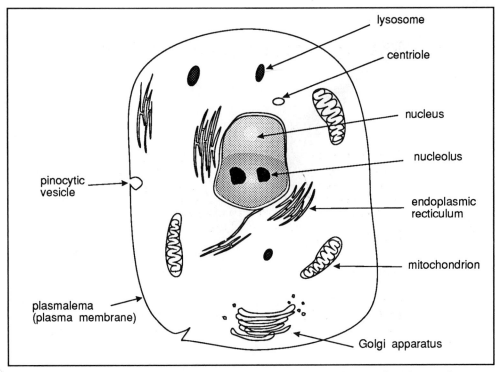

Figure 2.1 Structure of a generalised animal cell. The average animal cell is approximately 20μm in diameter.

cytoplasm

cytosol

mitochondria

endoplasmic reticulum

ribosomes

The compartment of the cell is divided into the nucleus and the cytoplasm. The cytoplasm consists of a number of structures called organelles, floating in an aqueous solution containing small molecular weight compounds, this solution being referred to as the cytosol. The organelles which often appear to be sausage shaped are the mitochondria (singular mitochondrion). Aerobic respiration, the process of oxidation of compounds to release their energy, is carried out in the mitochondria. The endoplasmic reticulum occurs as both rough and smooth forms, the rough form being due to the presence of densely-staining particles, the ribosomes. The presence or absence of ribosomes is not however the only difference between them. The smooth endoplasmic reticulum is much more variable in shape, being represented by circular and extended outlines, some of which appear to be branched. The two forms are very often found adjacent to one another, raising the possibility that they may be interconvertible. Note that ribosomes also occur free in the cytosol. These have not been shown in Figure 2.1 because although they are numerous, they are very small and can only be seen under very high magnification. The ribosomes are where proteins are synthesised. Free ribosomes produce proteins for use within the cell and those on the rough endoplasmic reticulum produce proteins for export out of the cell. Note that the ribosomes are present on only one side of the membrane of the endoplasmic reticulum, illustrated in Figure 2.2.

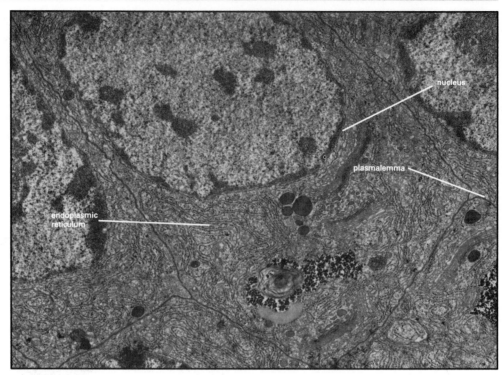

Figure 2.2 Electron micrograph of a thin section of an animal cell showing the rough endoplasmic reticulum (x 10 600).

Figures 2.1 and 2.2 show that the endoplasmic reticulum forms an internal space, which is separate from the rest of the cytosol. A further difference between rough and smooth endoplasmic reticulum is that the internal space is wider in the smooth than in the rough reticulum. In its function of exporting protein, the rough endoplasmic reticulum works in association with the Golgi apparatus and its vesicles. The smooth endoplasmic reticulum has the separate function of being responsible for lipid production. Also present are a number of filaments of various thickness which together form the cytoskeleton (Chapter 4). One group of these, the microtubules, are 25nm in diameter and are attached to another organelle, the centriole. The centriole functions in the process of cell division which is described in Chapter 8. Many animal cells also contain membrane bounded vesicles, the lysosomes, which contain digestive enzymes and which function in intracellular digestion.

Golgi

cytoskeleton

centriole

lysosomes

SAQ 2.1

What was the first component of eukaryotic cells to be discovered? Give reasons for your answers.

<table>
<tr><td>

SAQ 2.2

</td><td>

Which of the following are correct in relation to the statement 'prokaryotes differ from eukaryotes in the following ways?' Give reasons for your answers.

1) Prokaryotes have no internal membranes.

2) Prokaryotes have no organelles.

3) Prokaryotes are generally much smaller.

4) Prokaryotes show little internal specialisation.

</td></tr>
</table>

2.3 Plant cell structure

Figure 2.3 shows a stylised drawing of a leaf cell. Note that this cell contains a nucleus and cytoplasm containing organelles and thus shows eukaryotic cell organisation. There are several differences, however between plant and animal cells.

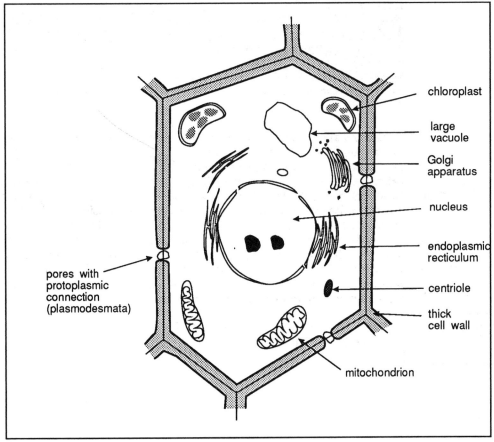

Figure 2.3 Structure of a generalised young plant cell. Note that in older, fully mature cells the vacuole often occupies up to 90% of the cell volume. In such cases the nucleus may be suspended (in the centre of the cell) by cytoplasmic threads.

SAQ 2.3	Using Figure 2.3 as a guide answer the following questions about plant and animal cells.

1) What organelles do animal and plant cells have in common?

2) Does the plant cell have any organelles not found in the animal cell?

3) Are there any other structural differences between plant and animal cells?

chloroplasts

vacuole

Figure 2.3 shows that in addition to the organelles found in animal cells, plant cells contain chloroplasts, which are approximately half the size of the nucleus. The process of photosynthesis is carried out in chloroplasts and starch is stored here. Plant cells also contain large vacuoles, which are relatively free of stainable contents, although they are often used to store toxic compounds. In fully grown plant cells, which may reach 100µm in diameter, the vacuole occupies more than 90% of the cell. In these cells, the nucleus is often situated in the middle of the vacuole suspended by cytoplasmic threads which link it to the remainder of the cytoplasm in the periphery of the cell.

cell wall

plasmodesmata

desmotubule

gap junctions

In addition, there is a cell wall surrounding the cell, completely enclosing it. The cell wall is made predominantly of cellulose fibres which form an interlinked network. Blotting paper is almost pure cellulose fibres. As with blotting paper, the spaces between the fibres in the cell wall can take up water and this is its usual state. This fluid forms the immediate environment of the cell, it is called extracellular fluid. Cell wall material is manufactured and exported by the Golgi apparatus. Spaces are also often present in the corners between cells. These are intercellular air spaces and are very important in the exchange of gases between plant cells and their environment. Note also the plasmodesmata. These appear as dark lines crossing the cell wall, linking the plasma membrane of one cell with that of the adjacent cell. Circular outlines are seen in glancing sections showing that plasmodesmata are tubular in nature. These circular outlines all contain a central core and this is considered to be a tubular piece of endoplasmic reticulum, called a desmotubule (Figure 2.4). Not all plasmodesmata contain desmotubules. The plasmodesmata are 20-40nm in diameter, large enough for viruses to pass through. This is how plant viruses are considered to move from one cell to another. Other evidence shows that metabolites of various sizes can easily diffuse through the plasmodesmata and this forms part of the communication system between cells and for the transport of nutrients and photosynthates from one part of the plant to another. Communication between animal cells is achieved using connections between cells called gap junctions. We will learn more about these when we learn about the structure of membranes in the next chapter.

2.4 Symplasm and apoplasm

symplasm

Plasmodesmata are found on all walls of plant cells forming a three dimensional system of linked cytoplasms. A name is given to this system: symplasm. By definition the symplasm is the cytoplasmic continuum of a plant tissue formed by the cytoplasm of adjacent cells linked up by plasmodesmata. It is a term describing part of a tissue, whereas the term cytoplasm describes part of a cell.

tonoplast
apoplasm

The vacuole is not part of the symplasm; compounds cannot move freely through its membrane, the tonoplast. So the tonoplast is the inner limit of the symplasm and the plasma membrane is its outer limit. The space outside the plasma membrane is called the apoplasm, which includes the cell wall and any space between it and the plasma

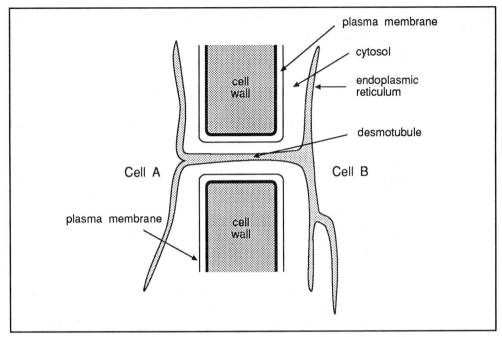

Figure 2.4 Structure of plasmodesmata.

membrane. The extracellular fluid occupies this space and because the gaps between the cellulose fibres are relatively large, compounds can more freely through the apoplasm. Compounds cannot, however, move freely between the symplasm and the apoplasm. The way movement between the two is achieved is described in the next chapter. Figure 2.5 illustrates the location of the apoplasm and symplasm.

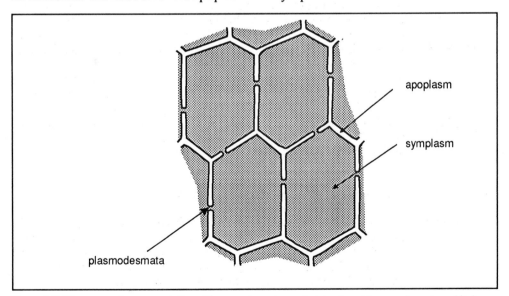

Figure 2.5 Diagrammatic representation of the symplasm (cross hatched) and the apoplasm (speckled). Only one plasmodesma is illustrated on each wall. The actual number varies between 10 and 50 and, of course, they are present on the 'front' and 'behind' face of the walls forming a three dimensional network.

| **SAQ 2.4** | Match the organelles structures listed below on the left with the function listed on the right. |

Organelle	Function
1) nucleus	a) intracellular digestion
2) chloroplast	b) protein synthesis
3) mitochondrion	c) plant cell wall manufacture and export
4) ribosome	d) cell division
5) rough endoplasmic reticulum	e) aerobic respiration
6) Golgi apparatus	f) controls cellular activities
7) centriole	g) protein synthesis for export
8) intercellular air spaces	h) photosynthesis
9) lysosomes	i) gas exchange

2.5 Interpretation of electron micrographs

It must be remembered that the micrograph is a two dimensional view of a three dimensional object and that a proper picture of the shape requires not only the sectioning of cells at different angles but also the use of serial sections. This approach has revealed, for example, that the endoplasmic reticulum forms a system of interlinked membrane-bound compartments (Figure 2.6).

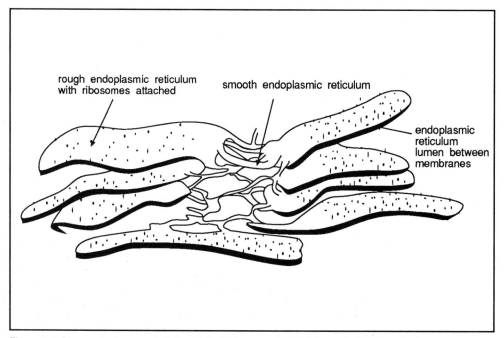

rough endoplasmic reticulum with ribosomes attached

smooth endoplasmic reticulum

endoplasmic reticulum lumen between membranes

Figure 2.6 Diagrammatic representation of the 3-D arrangement of the endoplasmic reticulum.

SAQ 2.5

The following shape is seen in an electron micrograph:

Which of the following solids could give rise to such a shape when sectioned?

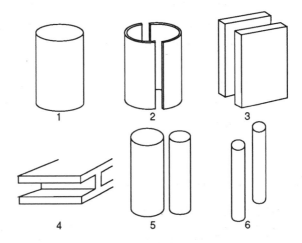

SAQ 2.6

Which of the following correctly states the meaning of the term organelle? Give reasons for your inclusions and exclusions.

1) Organelles are surrounded by membranes.

2) Organelles are used to store toxic compounds.

3) Organelles contain large insoluble compounds.

4) Organelles contain enzymes and components of all or part of a metabolic pathway.

2.6 The road to multicellularity

The major difference between prokaryotes and eukaryotes is fundamentally the presence or absence of internal membranes. The eukaryotic cell has lots of membranes and these are utilised to localise cellular activities in what are referred to as organelles. Thus instead of allowing all enzymes to be free to move anywhere within the cell those associated with one particular function, eg oxidative respiration, are concentrated in one type of organelle, in this case the mitochondrion. Prokaryotes do contain some internal membranes. Cyanobacteria for example contain internal membranes (thyllakoids) and the process of photosynthesis is catalysed by components localised on these membranes. Even so, in prokaryotes, this is the extent of locational specialisation and the enzymes catalysing other reactions tend not to be restricted in their distribution within the cell.

SAQ 2.7	Identify the structures arrowed on following micrograph (magnification = x 10 600):

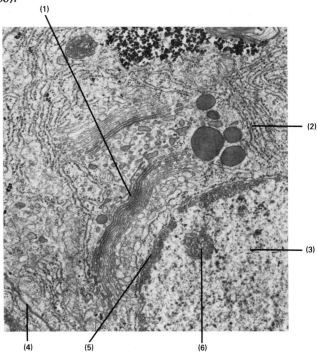

The internal organelle-based specialisation is considered to be central to the eukaryote cell's greater size. As a prokaryotic cell increases in size, enzymes become progressively more spread out thus reducing the speed at which long sequences of reactions can be completed. The consignment of all of the enzymes of a particular biochemical pathway into a particular organelle ensure its successful rapid operation. Thus we can understand, in part at least, how eukaryotic cells manage to be bigger.

∏   Why are eukaryotes capable of producing multicellular organisms?

It could perhaps be argued that the thick cell wall produced by bacteria would prevent cell to cell communication; something which must be essential for cells to function properly as part of a multicellular organism. Plants also have a thick cell wall but this has not prevented them from becoming multicellular. However plant cells have plasmodesmata which allow considerable cell to cell communication through the cell wall. Bacteria have no such connections and it seems reasonable to conclude that the very thick wall of the bacterial cell, while offering strength and protection, may have sentenced them to a life of solitude.

The matter is not however as simple as this. Let us consider some of the problems posed by multicellularity. In multicellular organisms we find different parts of each organism carry out quite different functions. Think about your own heart, lungs and kidneys or the roots, leaves and flowers of plants. We can readily recognise that there is considerable cellular specialisation in multicellular organisms. In other words, although cells from different parts of a multicellular organism have much in common, they each show some differences. For example, although chloroplasts have been described as organelles of plant cells, they are not found uniformly distributed. Root cells do not produce an abundance of chloroplasts! (Why not?) We call this specialisation of cells, cell differentiation.

specialisation

differentiation

The second issue of multicellular organisation is that the specialised (differentiated) cells are produced in specialised positions and in a co-ordinated manner. Our lungs, heart, kidneys and fingers are developed in a special spatial (ie not too big and in the right place) arrangement within our bodies. Likewise plant roots and leaves are also developed in functionally sensible arrangements. The development of the correct spacial arrangement we call morphogenesis.

morphogenesis

Cell differentiation and morphogenesis reinforce our observations that inter-cellular communication is vitally important. But they also imply something equally fundamental. These two features of multicellularity demand that considerable genetic information needs to be directed towards the regulation of these two processes.

genetic information

Let us make a brief contrast between unicellular and multicellular systems in terms of the genetic information they must carry. In a single-celled organism, the genetic information must contain all of the information necessary to make the enzymes and structures needed to:

- utilise available nutrients;

- synthesis cell components;

- carry out cell division.

In a multicellular eukaryotic system, the picture is much more complex. Multicellular systems usually begin life as a single cell (an ovum). This single cell must contain *all* of the genetic information that will be needed to specify all of the subsequent intracellular organelles and metabolism of all of its progeny. Thus, in humans, the fertilised egg cell must contain all of the genetic information that will be required to make the multitude of human cells (eg liver, kidney, red blood cells, white blood cells, skin cells, nerve cells etc). At the same time the fertilised egg cell must also contain the information which specifies which type of cell will be made in which position and how many of such cells will be produced. Although this is a somewhat simplified picture, the reader should be in no doubt that multicellularity demands much more genetic information than is required by unicellular systems and that much of this information is regulatory in nature.

The retention of greater quantities of genetic information imposes certain problems.

We have already learnt that the genetic information in eukaryotes is mainly retained in the nucleus. We will learn later that eukaryotes have developed a packaging system (chromosomes) which enable them to handle very large quantities of genetic information. Prokaryotes have not developed such a system. Many would argue therefore that the limited ability of prokaryotes to construct packaged genetic information means that their ability to carry and handle large amounts of genetic information is severely restricted. On this basis it is argued that the prokaryotes are restricted in their ability to develop multicellularity. If we couple this with the development of functional specialisation (ie organelles and chromosomes) at the sub-cellular level within eukaryotes, these may be the basis of cellular specialisation and the advent of multicellularity. With this in mind, we will explore in greater details the internal specialisation of eukaryotic cells before we go on to the questions of cellular specialisation and the 'division of labour' within multicellular systems. We begin by examining the properties of membranes since there is a key issue in understanding the structure and function of organelles.

Summary and objectives

In this chapter we have explained the general structural features of eukaryotic cells. We divided the discussion into plant and animal systems. Both types show a high degree of sub-cellular specialisation and we learnt how this specialisation is reflected in a greater degree of organization of metabolic processes. We also speculated about the importance and origins of multicellularity observed in eukaryotes. We were tempted to suggest that organelle specialisation led to cellular specialisation and the division of labour is characteristic of multicellular organisms. Now that you have completed this chapter you should be able to:

- describe what is meant by the term organelle;

- identify and state the function of the following organelles: the nucleus, ribosomes and endoplasmic reticulum, mitochondria, chloroplasts, the Golgi apparatus, cytoplasmic filaments and the centriole;

- show how two dimensional structures seen in an electron micrograph can be interpreted as three dimensional objects;

- show an understanding of the differences between eukaryotic and prokaryotic organisation;

- list similarities and differences between animal and plant cells;

- describe and recognise plasmodesmata and define the terms symplasm and apoplasm.

The structure and function of membranes

The structure and function of membranes

3.1 Introduction

Cells in multicellular organisms are maintained as distinct entities by the plasma membrane. This acts both as a barrier, separating the cell's internal solutions from those around it, and as a transport system, allowing the passage of certain compounds into or out of the cell. The functioning of this membrane is of paramount importance to the organism as a whole because the specialised functioning of cells depends upon the proper regulation of what goes in and what goes out. If cells are treated with chemicals which alter the permeability properties of the plasma membrane they can no longer maintain the correct internal environment and cease to function properly. In this chapter we will explore the structure of the plasma membrane, and indirectly, that of other cellular membranes, to see if we can understand how it carries out its various functions.

3.2 Early studies of the membrane needed a readily available source of cells

If a scientist is planning to study plasma membranes a suitable source of material is necessary. One particular source has been used much more than any other.

∏ Can you suggest what source of plasma membranes have been used more than any other? Would it be of plant or animal origin?

Ideally this source will be readily available and it must be reasonably easy to obtain plasma membrane material from it. If the plasma membrane can be obtained in relatively pure form so much the better.

∏ Do you know of a cell type which satisfies these criteria?

The answer is the human red blood cell. This is a quite unusual cell type because when it is fully mature it no longer contains a nucleus and has virtually no other internal membranes. It is also available from blood banks in considerable quantity. Thus red blood cells are a potentially useful source of membranes.

Having got the cells, we now break them open by lysis, using the process of osmosis. For this we place them in a dilute solution. The cells take up water by osmosis and swell. Eventually they rupture and the contents of the cells are released, leaving behind the cells' membranes. The mixture is centrifuged so that we finish with the red cell membranes concentrated at the bottom of the tube. These are referred to as red cell

red cell ghosts   ghosts. Chemical analysis of red cell ghosts shows that they contain lipid and protein with a small amount of carbohydrate. This much was known by the early 1920s and it was soon after this, in 1925, that a very important experiment was done. Lipid was extracted from a known quantity of red cells and floated on the surface of water. The area of the lipid layer was decreased by a movable barrier until it formed a monolayer, a layer only one molecule thick. The area of the monolayer was measured and found to be almost exactly twice that of the red cells from which it came.

Π   What do you think this might suggest about the arrangement of lipid in the plasma membrane of the red cells?

bilayer

The most likely explanation is that the lipid exists as a bilayer. This has been confirmed using X-ray diffraction, a technique which gives physical information about the shape and distribution of parts of molecules.

amphipathic

hydrophilic

hydrophobic

The lipids found in membranes consist of phospholipid, cholesterol and glycolipids. These have in common the fact that they are amphipathic, which means that they have a hydrophilic (water liking or polar) end and a hydrophobic (water hating or non polar end). We have included a brief description of these molecules in Chapter 1. At that time no clear idea was available as to how the protein fitted in with the lipid but that was soon to change.

SAQ 3.1

1) If the structure of membrane lipids is represented like a match with the head representing the polar end, and the stick representing the non-polar end, which of the two arrangements shown in Figure 3.1 would you expect to find in the plasma membrane? Give your reasons.

2) Under what conditions would the other arrangement occur?

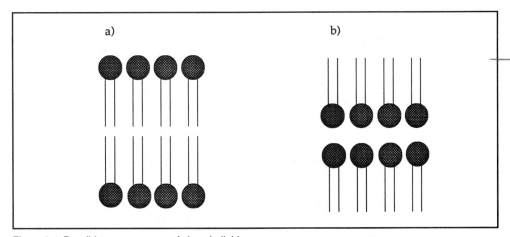

Figure 3.1 Possible arrangements of phospholipids.

3.3 The electron microscope reveals some of the structure of the membrane

The electron microscope enables thin sections of tissues to be examined at much higher magnification than the light microscope and under it the plasma membrane shows a three-layered structure (Figure 3.2).

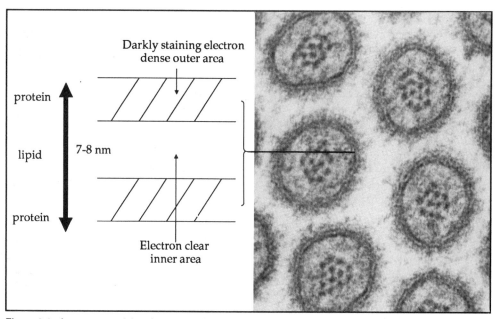

Figure 3.2   Appearance of the plasma membrane in the electron microscope. The electron micrograph is of a transverse section of microvilli of intestinal cells (x 200 000).

unit membrane

Other membranes in the cell also show this three-layered sandwich raising the possibility that they all have the same basic structure. Note that the thickness of these membranes is 7-8nm. The name unit membrane was coined for this universal structure.

position of proteins

The clear portion of the three layered sandwich was considered to be a phospholipid bilayer and the two dark lines are formed by two layers of protein one on each side of the lipid. This might seem a logical arrangement because the hydrophobic part of the lipid is buried deep in the centre while the polar ends associate with the protein which, it was assumed, is polar too. This is not completely true because some of the amino acid residues of the protein are hydrophobic and would not readily associate with the polar ends of phospholipids. However, the exact mode of association of the protein with the lipid was never clearly spelled out in the unit membrane theory and this was in fact a drawback. Before examining the next advance in our understanding of membrane structure, let us examine certain properties of the membrane so that we know what we have to be able to explain.

3.4 The membrane controls what goes into and what comes out of cells

We have already stated that the plasma membrane isolates the cell contents from the surrounding solution and also controls what goes in and what goes out. Let us examine the concentration of compounds inside and outside cells to get a clear picture of what is involved.

Component	Intracellular concentration (mmol/litre)	Extracellular concentration (mmol/litre)
Na^+	10	145
K^+	140	5
Ca^{++}	1-2	2.5
Mg^{++}	30	1.5
glucose	5.5	5.5

Table 3.1 Comparison of solute concentrations inside and outside a typical mammalian cell.

All of the compounds listed inside the cell in Table 3.1 were once in the extracellular fluid and, therefore, must have passed through the plasma membrane to get into the cell. The simplest way for this to occur is by the process of diffusion, in which compounds migrate down gradients of concentration until they are uniformly distributed. At this point the compounds are moving at the same rate in both directions.

∏ Can the concentrations of all of the intracellular compounds listed in Table 3.1 be explained by simple diffusion into the cell?

The answer is no. Look, for example, at the difference in concentration of the potassium ions (K^+) inside and outside of the cell. We will return to this point later.

3.5 Osmosis is a special type of diffusion

osmosis

lysis

We normally think of the diffusion of solutes, but the solvent can also diffuse and it, too, moves from a region of high to one of low concentration until the concentrations are equal. Diffusion of a solvent through a membrane which tends to hold back solutes but not the solvent (ie it is semipermeable) is called osmosis. In biology, water is the universal solvent so when biologists talk of osmosis they are referring to the diffusion of water through a semi-permeable membrane. The plasma membrane is semi-permeable so water will enter the cell by osmosis if the concentration of water is higher outside than inside. This increases the volume of the cell which will burst if osmosis is allowed to proceed for long enough. We have already learnt that red blood cells can be caused to burst (to lyse), by placing them in water. Plant cells do not burst under these conditions because their cellulose cell walls prevent it.

isotonic

hypotonic

hypertonic

Solutions with the same water concentration as each other are said to be isotonic. A solution with a higher water concentration than another is said to be hypotonic to the other. A solution with a lower water concentration than another is said to be hypertonic to the other. If cells are placed in a hypertonic solution the cell shrinks because water moves out of the cell by osmosis. This is not a very striking phenomenon in animal cells but it can be readily observed in plants. The shrinking protoplasm no longer fills the space inside its cell wall and this can be seen clearly with the light microscope. A stylised picture of this is given in Figure 3.3.

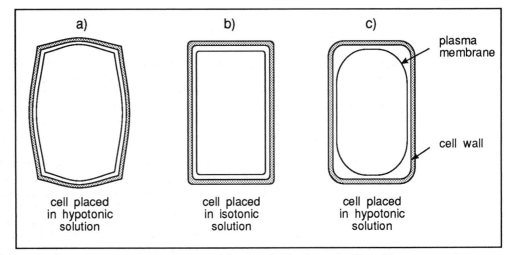

Figure 3.3 Effect of osmosis on the volume of the protoplast in plant cells. The cell loses water to its environment in c) but gains water in a).

plasmolysis

turgid

The cell in Figure 3.3c is said to be plasmolysed and is characteristic of tissue which is showing wilting, such as a limp lettuce leaf. The cell in Figure 3.3a is turgid and is characteristic of a crisp lettuce leaf. The turgidity of plant cells is important in helping maintain leaves in their normal orientation.

SAQ 3.2

If plants are exposed to conditions which cause extremes of plasmolysis for long periods they may not survive. Can you think of a reason for this? (Hint, think about proteins).

3.6 The transport of some compounds needs energy

active transport

Examine Table 3.1 again. You will note that some components are present at a higher concentration inside the cell than outside; ie they have been absorbed against a gradient of concentration. This cannot occur simply by process of diffusion for this would lead to concentrations being equal inside and outside the cell. Movement of a compound across a membrane may be influenced by charge on the membrane and the charge inside and outside of the cell. Frequently, however, movement of compounds against a concentration gradient is achieved by an energy demanding process called active transport. In this process ATP is used to drive the movoement of the compound against the concentration gradient.

SAQ 3.3   Which of the compounds shown in Table 3.1. look as if they have been absorbed by active transport and which by diffusion?

In Table 3.1 not all of the extracellular components have been absorbed against a concentration gradient; the process is selective. This is the essence of the function of the plasma membrane. It can transport *selected* compounds against a concentration gradient. Only with this quality of selectivity in its plasma membrane can cellular integrity be maintained.

How is cellular integrity achieved?

Experiments using artificial membranes made of lipid bilayers show that they are freely permeable to certain small uncharged compounds such as water, carbon dioxide and oxygen, moderately permeable to larger uncharged molecules such as glucose and sucrose but virtually impermeable to compounds carrying a net charge even if they are as small as sodium or potassium ions. The plasma membrane of intact cells behaves similarly to the artificial membrane but there are some important differences.

proteins as aids to trans-membrane transport

As we have seen in Table 3.1, potassium ions are capable of penetrating plasma membranes and experiments in which this process is carefully monitored show that it has similarities to an enzyme-catalysed reaction. It is possible for example to determine Km and Vmax values. (Km is a measure of the affinity of an enzyme for its substrate. Vmax is a measure of the maximum velocity of an enzyme catalysed reaction - further discussion of these properties are given in the BIOTOL text 'The Molecular Fabric of Cells'). This raises the possibility that factors in the membrane itself are behaving as enzymes, their function being to catalyse the movement of compounds through the membrane. It is tempting to suggest that proteins in the membrane are responsible for this. If the unit membrane structure is correct, however, some way must be found to transport the ion across the barrier of the lipid bilayer. For this reason it was proposed that pores were present in the membrane, lined by protein, which allowed the transport of lipid-insoluble compounds. We briefly discussed the role of proteins in the transport of materials across the plasma membrane in prokaryotes in Chapter 1. We did not however discuss the structural evidence for the model we drew. In the next section we explain some of the evidence that supports this model.

3.7 Freeze-fracturing provides a new view of the membrane

The methods discussed so far that have been used to study cell structure are not those which would produce three dimensional images directly and to obtain such pictures painstaking serial sectioning is required. The development of the freeze-fracturing technique provided 3-D images directly for the first time.

In this technique the cell is rapidly frozen in liquid nitrogen to a temperature below -100°C and then cut or fractured with a microtome in a cooled chamber. The knife fractures cells along zones of weakness. In a frozen tissue these are along the surface of membranes and, more importantly for our purposes, along the hydrophobic centres of membranes. Hydrophobic attractions are among the weakest of those holding molecules together and in the frozen state they are even weaker. Thus freeze-fracturing breaks membranes open by splitting them down the middle and allows us to look at the exposed surfaces. The steps in this procedure are illustrated in Figure 3.4.

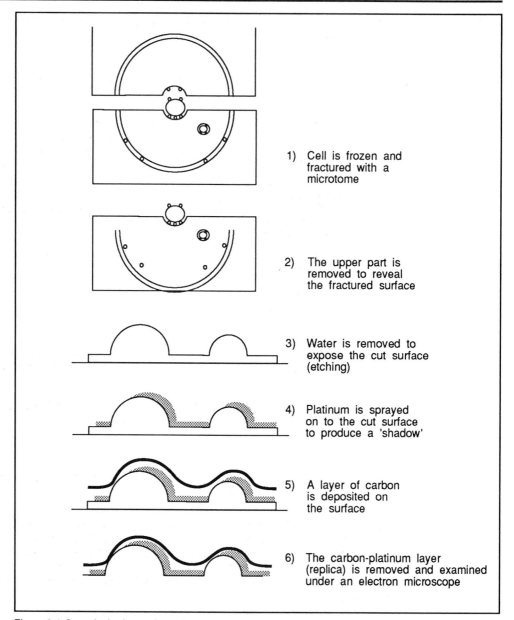

1) Cell is frozen and fractured with a microtome

2) The upper part is removed to reveal the fractured surface

3) Water is removed to expose the cut surface (etching)

4) Platinum is sprayed on to the cut surface to produce a 'shadow'

5) A layer of carbon is deposited on the surface

6) The carbon-platinum layer (replica) is removed and examined under an electron microscope

Figure 3.4 Steps in the freeze-fracturing process (stylised).

Once fractured the exposed surface is etched to remove water and to reveal organelles and then shadowed with platinum by spraying it obliquely onto the surface. A layer of carbon is sprayed onto the surface creating a replica of it. This can then be lifted off with the platinum layer and be examined with the electron microscope.

∏   What is the purpose of shadowing with platinum?

Platinum is electron dense. Platinum shadowing therefore increases the contrast and makes the relief of the surface more marked.

∏ Why is a carbon replica used rather than the cut cell itself? (Think about the thickness of the specimen being examined).

Because the cell is relatively thick, electrons will not pass through it. Thus the cell will appear as a totally black object in the electron microscope. Electrons can pass through the replica and thus this can be viewed with the electron microscope.

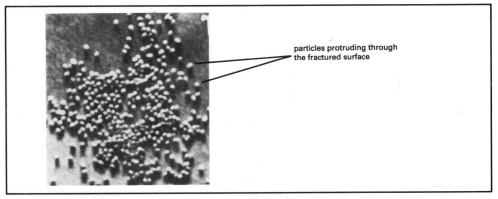

particles protruding through
the fractured surface

Figure 3.5 Face-view appearance of red cell plasma membranes revealed by freeze-fracturing.

Figure 3.5 shows the result of freeze-fracturing red blood cells in such a way as to fracture the plasma membrane. Note the particles which are present over all of the preparation shown.

∏ What is the nature of the particles shown in Figure 3.5? How would you prove your answer was correct?

Figure 3.6 shows what happens when the fractured surfaces shown in Figure 3.5 are treated with pronase, a proteolytic enzyme, prior to producing the replica. Note that the density of particles gets less and less with increasing time of treatment suggesting that the particles are protein in nature. These results are interpreted as suggesting that proteins are present amongst the phospholipid of the membrane and not just on its outer surface.

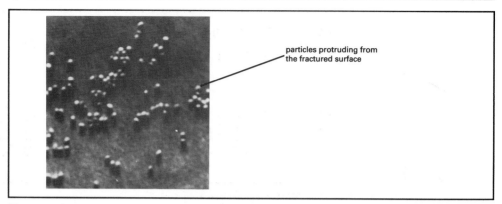

Figure 3.6 The effects of pronase on red blood cell plasma membrane. The plasma membrane has been exposed to pronase. Compare this figure with Figure 3.6.

fluid-mosaic
model

It was also realised that the lipids found in membranes would be liquid at normal temperatures.These observations led to the proposal of the Fluid-Mosaic model of membrane structure, in which protein and lipid form a mosaic, as viewed from above, and that proteins penetrate completely through the lipid bilayer (Figure 3.7).

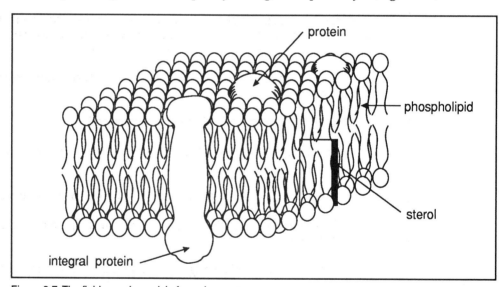

Figure 3.7 The fluid-mosaic model of membrane structure.

SAQ 3.4

Figure 3.8 shows a stylised representation of a protein embedded in a lipid bilayer. What properties can you suggest that the shaded and non-shaded portions of the protein may have?

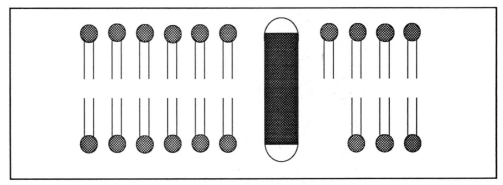

Figure 3.8 A protein embedded in a lipid bilayer (stylised).

<table>
<tr><td>

SAQ 3.5

</td><td>

It is possible to label membrane proteins with a fluorescent dye thereby enabling them to be seen in a fluorescence microscope. A number of dyes are available which can be visually distinguished from each other. Examine the experiment described below and answer the questions.

Membrane proteins of human cells and mouse cells were labelled with two different fluorescent dyes. The two cells were allowed to fuse and the membrane proteins visualised giving the results in Figure 3.9.

Note that the membrane proteins migrate after fusion and become uniformly distributed.

</td></tr>
</table>

1) Does a) or b) better explain this phenomenon?

   a) proteins in a membrane are part of a static structure like bricks in a wall.

   b) proteins in a membrane are highly mobile in a two dimensional fluid.

2) What would be the effect of an increase in temperature on this process of migration?

3) How could you explain migration stopping at a low temperature?

integral
and
peripheral
proteins

In addition to the proteins embedded in the lipid bilayer we now know that there is a second group of membrane proteins, attached to the outer portions of the embedded proteins and associated with the outer surface. In order to distinguish these, the embedded ones are called integral membrane proteins and the surface ones peripheral membrane proteins. To complete the picture; you may remember early in this chapter we noted that small amounts of carbohydrate are found in membranes. Studies show that small numbers of sugar residues forming oligosaccharides are attached to protein and to lipid forming glycoprotein and glycolipid, respectively. These carbohydrates are present, however, only on the outer surface of the plasma membrane, forming what is called the glycocalyx.

glycocalyx

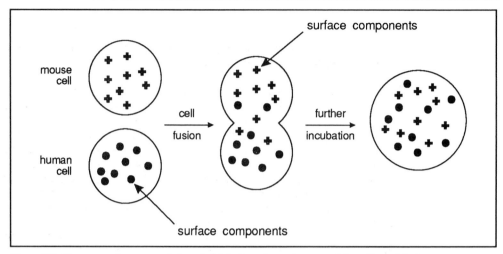

Figure 3.9 Diagrammatic representation of distribution of membrane proteins after cell fusion.

Some of the integral proteins are considered to function as carriers (transport enzymes) and to catalyse the movement of certain compounds against a concentration gradient. In this process ATP is hydrolysed. We can represent this process diagrammatically by:

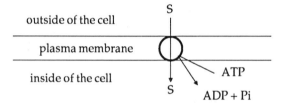

Several different mechanisms are thought to operate. This is currently a topic of very active research.

3.8 Facilitated diffusion is a rapid form of diffusion

As you can imagine it takes a certain amount of time for a compound to diffuse into a cell until its concentration inside is equal to that outside. A number of examples have been found, however, where this occurs more quickly than expected, but the overall process still has the basic property of diffusion. No ATP is consumed in these examples so the process is not active transport. The name facilitated diffusion is used for this. The absorption of glucose by red blood cells is an example. It is considered to occur by reaction of glucose with an integral protein which somehow aids its movement through the membrane. This protein can be considered to be a carrier enzyme but it does not need ATP for any part of its function. In discussing the plasma membranes of prokaryotic cells, we mentioned the presence of some of these carrier proteins (eg porins).

facilitated
diffusion

3.9 Gap junctions allow the passage of small molecules and ions between adjacent cells

gap junctions

Experiments conducted for the first time in 1958 demonstrated that certain animal cells could exchange inorganic ions relatively freely from one cell interior to the other. This demonstration was achieved by inserting micro-electrodes into adjacent nerve cells and showing the passage of a current between them. Inorganic ions, which carry current in living tissues, must be moving from one cell to the next. This phenomenon was later confirmed using fluorescent compounds. These compounds were injected into one cell and were soon detected in the next, but not in the space between them. The use of compounds of increasing molecular mass showed that molecules smaller than 1500 daltons could pass through, indicating a pore size of between 1.5 and 2.0 nm. This is about one tenth the diameter of the plasmodesmata which link adjacent plant cells. The pores between animal cells are called gap junctions and they have been found in most tissues examined. Their structure has been elucidated by freeze-fracture studies (Figure 3.10).

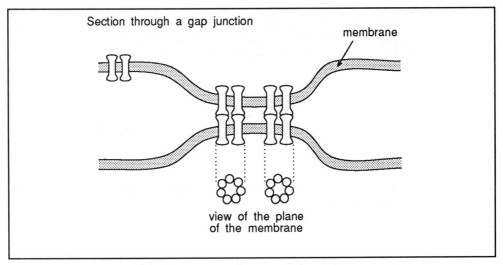

Figure 3.10 Structure of a gap junction.

connexons

The basic unit of a gap junction is an integral protein which protrudes on each side of the membrane; it is called a connexon. Connexons can migrate laterally in the membrane and when connexons of adjacent cells align with each other they bind and form the double connexon, which has a pore running through it linking the cytosol of the adjacent cells. Presumably the pore in single connexons is closed otherwise the cells would be leaky. The formation of the double connexons forms a junction between cells but there is a gap between them of 2-4nm; hence the name gap junction. Gap junctions typically contain many hundred connexons but they can be formed or disconnected very quickly by the lateral migrations of individual connexons.

gap junction and plasmo-desmata

Although the evidence is only circumstantial, gap junctions appear to allow metabolic cooperation between cells suggesting that they are analogous to the plasmodesmata of plant cells. The most striking illustration for this is the synchronous differentiation of groups of animal cells during their development. This appears to depend on the speedy distribution of morphogenetically-active compounds between the cells of the group. However, it has not yet been possible to reconstitute functional gap junctions in artificial membranes, which would be a considerable step in the unambiguous demonstration that molecules can pass through them.

3.10 The transport of large molecules involves vesicles

exocytosis

In addition to the small molecular mass compounds examined so far, cells can also take in or eject large molecules such as proteins and polysaccharides. This involves a quite different mechanism from those described above. Large molecules, such as insulin, are packaged into small membrane-bound vesicles which then fuse with the plasma membrane and release their contents into the extracellular space. This process is called exocytosis. It is shown diagrammatically in Figure 3.11 along with two other related processes, endocytosis and budding.

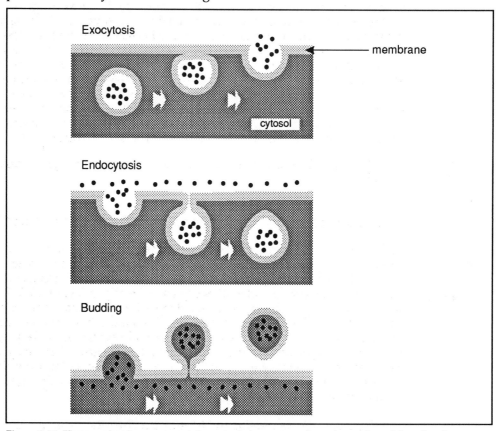

Figure 3.11 Diagram showing sequence of events during exocytosis, endocytosis and the related phenomenon, budding. Note that membrane proteins have been omitted for simplicity.

SAQ 3.6

It has been said that exocytosis is simply the reverse of endocytosis. Using Figure 3.11 as a guide can you suggest at least one reason why this can not be so?

SAQ 3.7

Having answered SAQ 3.6 and checked the response, you should be able to suggest a role for membrane proteins in the membrane fusion processes of exocytosis, endocytosis and budding.

SAQ 3.8 Examine the sequence of events of budding in Figure 3.11. The implication from the diagram may be that budding only occurs into the external environment of a cell. This is not necessarily the case. Can you suggest a situation inside the cell where budding might occur?

3.11 Macromolecular secretion occurs by exocytosis

triggered
exocytosis

Most animal cells secrete macromolecules into their environment using the process of exocytosis. Some of the molecules released stay attached to the outer membrane surface while others diffuse into the extracellular fluid. Some serve as nutrients and others as signals to other cells. Many vesicles are continuously secreted but those used as signals to other cells are released at particular times. This is called triggered exocytosis, the release being activated on receipt of a signal from another cell. This signal is very often in the form of a chemical message, such as a hormone. An important point to realise in all cases of exocytosis is that vesicle membranes become incorporated into the plasma membrane and this leads to an enormous increase in the surface area of the cell. The acinar cells in the pancreas produce food digestive enzymes which are secreted by triggered exocytosis through the portion of the cell, the apical surface, which faces the secretory tubule. On receipt of the signal about $900\mu m^2$ of vesicle membrane is added to a surface which is normally only $30\mu m^2$ in area. This extra membrane is retrieved by endocytosis and the normal membrane surface area is re-established.

constitutive
exocytosis

Exocytosis, which occurs continually and does not need a trigger, is called constitutive exocytosis.

3.12 Macromolecular absorption occurs by endocytosis

pinocytosis

phagocytosis

Technically, two types of endocytosis are recognised, according to the size of particle ingested. Pinocytosis (literally 'cell drinking') is used when small macromolecules and fluid are taken in. Large particles (eg micro-organisms or cell debris) can be ingested in the form of endocytosis called phagocytosis (literally 'cell eating'). While virtually all animal cells carry out pinocytosis (for them it is constitutive), phagocytosis is carried out by specialised cells called phagocytes. Most endocytotic vesicles fuse with primary lysosomes (the sites of cellular digestion) and become secondary lysosomes. The products of digestion are transported into the cytosol where they are used. The membrane of the endocytotic vesicle is in some way protected from digestion and is retrieved for further use, probably by budding from the secondary lysosome followed by incorporation into the plasma membrane.

Not all endocytotic vesicles fuse with lysosomes. In certain specialised cells, vesicles are produced by ingestion (endocytosis) on one side of the cell, transported across to the other and released by exocytosis. This method of bulk transcellular transport is one of the ways in which the endothelial cells that line small blood vessels transport compounds from the blood stream to the extracellular fluid.

Another example of endocytotic vesicles which do not fuse with lysosomes is receptor-mediated endocytosis. In this process the exogenous agent binds with particular cell surface components (receptors). The cell surface invaginates to form a vesicle. Molecules can be secreted from a cell by a reversal of this process (ie receptor-mediated exocytosis).

3.13 Movement of large molecules also occurs in plant cells

Plant cells rarely if ever ingest large molecules from their environment and, therefore, do not often exhibit endocytosis. They do show exocytosis, however, but this is limited to cells which are growing. The raw materials required to build the cell wall are manufactured inside the cell and secreted into the extracellular space by exocytosis. Once a cell is mature and carrying out its normal function exocytosis does not seem to be part of its everyday activities. We noted earlier that the plant cell wall prevents the undue increase in cell volume which might otherwise occur as a result of water absorption by osmosis. This results in the plasma membrane being pressed hard against the wall. It is difficult to see how endocytosis could occur because the internal pressure would be forcing it out all the time. Exocytosis is perhaps easier to understand under these circumstances since it is likely that the plasma membrane is actually stretched. The insertion of a portion of extra membrane would partially relieve this tension and presumably would occur almost spontaneously.

SAQ 3.9

Signify with a cross in the appropriate place below, if the statements apply to exocytosis and/or endocytosis.

1) Requires an extracellular signal.

2) Requires fusion of non-cytosolic faces of membrane.

3) Requires fusion of cytosolic faces of membrane.

4) Produces a vesicle which fuses with a lysosome.

5) Produces a vesicle which does not fuse with a lysosome.

6) The term constitutive may apply.

7) Is involved with transporting materials across cells which line blood vessels.

8) Is involved with transporting materials in plant cells.

	Exocytosis	Endocytosis
1)		
2)		
3)		
4)		
5)		
6)		
7)		

Summary and objectives

In this chapter we have learnt that the plasma membrane of a cell separates the cell's contents from the cell's environment and controls the transfer of materials between the two. Its proper functioning is crucial to the life of the cell.

Using the data derived from studies on red blood cell ghosts and electron micrographs, we explored how ideas on membrane structure evolved through the Unit Membrane model to the currently accepted Fluid Mosaic model. We have also learnt that membranes selectively allow the passage of small solute molecules and we have explored the consequences of osmosis on cell function and appearance. We have considered the role of membranes in the passage of macromolecules and larger biological structures. We learnt that uptake can occur by endocytosis and that excretion can occur by exocytosis. We have also learnt that endocytosis can be divided into pinocytosis and phagocytosis and that these processes are mainly confined to animal systems. We have learnt that the passage of materials into and out of the cells, may be by diffusion, by facilitated diffusion or by the energy demanding process of active transport.

Now that you have completed this chapter, you should be therefore able to:

- distinguish between diffusion and active transport and deduce which has occurred from evidence provided;

- explain the meaning of the terms osmosis and facilitated diffusion;

- explain what is meant by freeze-fracturing, and use presented data to suggest models of membrane structure;

- correctly interpret presented data related to the fluid nature of the plasma membrane;

- show an understanding of the terms exocytosis and endocytosis and give examples of various types of each;

- define gap junctions and describe evidence for cell to cell transport in animal cells.

The maintenance of cell shape: the cytoskeleton and the plant cell wall

The maintenance of cell shape: the cytoskeleton and the plant cell wall

4.1 Introduction

Eukaryotes are made of large numbers of types of cells, usually specialised for a given function and many of these can be recognised by their shape (Figure 4.1). In animals they include the striking and unusual nerve cells, where one part of them, the axon is a process which may be several metres long. Bone cells are more regularly-shaped cells, but with numerous extended processes which connect to similar ones on adjacent cells. Smooth muscle cells are narrow with pointed ends. In addition, cells such as the fibroblast cells of connective tissue can change their shape and also migrate from one place to another. The plant world contains almost as much diversity. The longest plant cell is undoubtedly the pollen tube cell, which can be 1-2cm long and plants also contain narrow pointed end cells; the tracheids and fibres. The largest xylem vessel cells are so wide that they can almost be seen with the naked eye. Look at the examples shown in Figure 4.1, they represent only a tiny fraction of the many hundreds of different cell types found in multicellular organisms. They are different because they carry out different functions (ie there is a 'division of labour'). This specialisation of cells, or cell differentiation as it is called, is the focus of Chapter 9. We have seen in the previous chapter that the plasma membrane consists essentially of protein molecules floating in liquid phospholipid. It is difficult to see how a liquid limiting membrane can exert sufficient force to produce anything other than a spherical blob, shaped only by being pressed against adjacent cells. The shape of plant cells is obviously governed by the cell wall; once this is formed it is easy to see how shape is maintained. However, we still have to explain how the shape is formed in the first place. Animal cells, of course, have no equivalent to the plant cell wall.

It is now known that the property of cell shape and movement depend on a complex network of interlinked protein filaments which has been given the name cytoskeleton. This is a slightly misleading name for it implies a static structure, whereas the cytoskeleton is flexible, many of its elements being in a highly dynamic state and constantly changing. The filaments which make up the cytoskeleton were originally classified on the basis of their relative sizes. Microtubules are the broadest and are 25nm in diameter, whereas microfilaments, now referred to as actin filaments, are 7nm in diameter. For a while no other filaments were evident but then others were discovered which were intermediate in diameter between microtubules and actin filaments. They are known as intermediate filaments. Their diameters range from 8-10nm. Whereas microtubules can be seen easily in electron micrographs, actin and intermediate filaments cannot and are often grouped as cytoplasmic filaments.

cytoskeleton

microtubules

microfilaments

actin

intermediate filaments

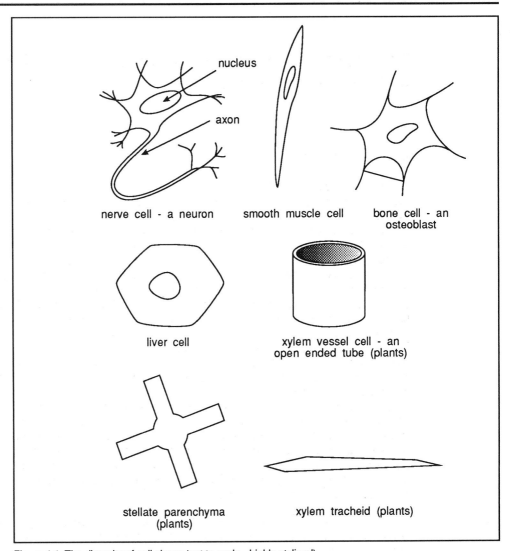

Figure 4.1 The diversity of cell shape (not to scale - highly stylised).

In this chapter we will first describe the structure and function of the filaments which make up the cytoskeleton in animal cells. The structure of plant cell walls will then be described and the role of the cytoskeleton in the control of cell orientation discussed.

| **SAQ 4.1** | Which of the following apply to the cytoskeleton?

1) It is rigid and gives strength to a cell wall.

2) It is always present and never changes.

3) It forms an interlinked framework of filaments.

4) The cytoskeleton can not be seen in the electron microscope.

5) Intermediate filaments are so called because they link microtubules and actin filaments.

4.2 Microtubule structure

tubulin   Microtubules are hollow rod-shaped organelles made of tubulin, a globular protein. Two forms of tubulin are found, α and β tubulin, both with a molecular mass of 50,000 dalton and both having guanine 5'-triphosphate (GTP) bound to them. These monomers associate in pairs to form tubulin dimers which are arranged in a helix around a central core which appears empty in electron micrographs (Figure 4.2).

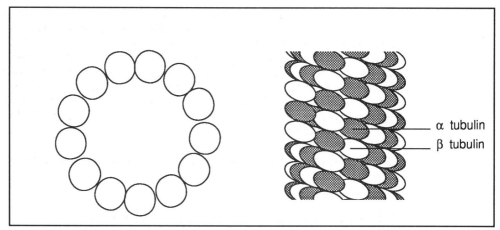

Figure 4.2 Structure and assembly of microtubules. Note that in a vertical column the α tubulin of one dimer is adjacent to the β tubulin of the next.

associated   The assembly of tubulin molecules into microtubules occurs spontaneously *in vitro* and
proteins   is normally accompanied by the hydrolysis of GTP to GDP and phosphate. Many microtubules, however, are labile and indeed their function sometimes depends on this liability. A very striking example of this is the feeding tentacle of certain heliozoans eg *Actinosphaerium*. During feeding the tentacle attaches itself to prey and then rapidly retracts to the cell body bringing the prey with it. The tentacle contains hundreds of microtubules running longitudinally in it and their rapid depolymerisation causes it to retract. With such groups of microtubules arranged in parallel fashion there is a very strong evidence that accessory proteins form lateral bridges between adjacent microtubules. These are called microtubule-associated proteins.

Two types of drugs have been found to be of use in the study of microtubule assembly and disassembly.

colchicine   Colchicine is an alkaloid that has been known since ancient Egyptian times as a treatment for gout. Colchicine molecules bind on a one to one basis to the tubulin dimers and prevent their polymerisation. Thus addition of this drug results in the disappearance of microtubules because assembly is prevented. The application of this drug can be used to diagnose those systems which are dependent on labile microtubules. The microtubules of the spindle, used to separate chromatids during mitosis (see Chapter 8), are such a system and colchicine inhibits this process and prevents cell division. Colchicine and similar drugs, because of their anti-mitotic activity, are widely used as anti-cancer agents because the disruption of mitotic spindle

taxol   microtubules preferentially kills rapidly dividing cells. A different type of drug, taxol, does the opposite in that it reduces disassembly and stabilises the microtubule. Interestingly this drug also inhibits mitosis. Mitosis is dependent, therefore, on microtubules being in a constant state of assembly and disassembly. This (as we shall see later) is not the case for all systems containing microtubules.

4.3 Microtubules can grow at either end

Tubulin dimers can be added to and removed from either end of a microtubule but the two rates are not the same when GTP is present. If we label one end of the microtubule as the (+) end and the other as the (-) end, then in the presence of GTP the (-) end tends to disassemble while assembly continues at the (+) end. At certain concentrations of tubulin, at which the overall rate of addition of dimers equals the overall rate of removal, there will be a net movement of dimers from the (-) end to the (+) end. The

treadmilling   microtubule will stay the same length. This is called treadmilling and, although it has not yet been demonstrated *in vivo*, it could provide a mechanism whereby objects can be pushed or pulled inside the cell. At the low concentration of tubulin found in cells it might be expected that the rate of removal from the (-) would be greater than the rate of incorporation at the (+) end. Thus a microtubule would eventually disappear unless its (-) end was protected (Figure 4.3).

The occurrence of microtubules in cells implies that they are protected at least for part of the time.

∏   How is the protection of microtubules brought about?

centrosome   Microtubules are in fact found to grow from certain regions called organising centres which protect their (-) end. These regions in animal cells are called the cell centre or centrosome and animal cells are often found to have a centriole pair at their centre. Each centriole half-pair is a cylinder 0.1μm in diameter and 0.3μm long, the two cylinders of each half pair being at right angles to each other. Each cylinder has been found to contain a ninefold array of microtubule triplets (Figure 4.4) and the possibility is raised that the centriole is the microtubule organising centre. Evidence in favour of this comes from the study of the structure of cilia and flagella.

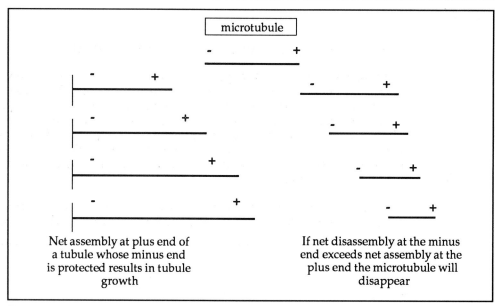

Figure 4.3 Model showing the fate of free and protected microtubules at the low tubulin concentration found in cells.

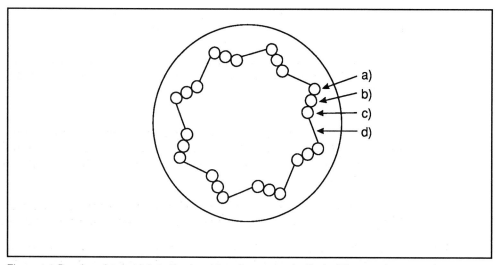

Figure 4.4 Drawing of a centriole half-pair sectioned transversely. Each triplet contains a complete microtubule a) plus two incomplete ones (b and c). Triplets are held together by accessory protein d).

4.4 Microtubules occur in cilia and flagella

Cilia are fine hair-like appendages, about 0.25μm in diameter and 20μm in length, which extend from the surface of many cells such as those which line the respiratory passageways, the trachea and tracheoles. Cilia are generally used to propel fluid over the cells surface but in some unicellular organisms, such as *Paramecium*, they occur over all the surface and are used to propel the organism itself. A flagellum is similar to a cilium but many times longer. They usually occur singly. The beating of flagella is more commonly used to propel an organism such as the green alga *Chlamydomonas*. It is also the locomotory system used by sperm.

Paramecium

Chlamy-domonas

The interesting thing about flagella and cilia, for our purposes, is that they are basically made of microtubules. Each contains 11 microtubules in a 9:2 arrangement (Figure 4.5).

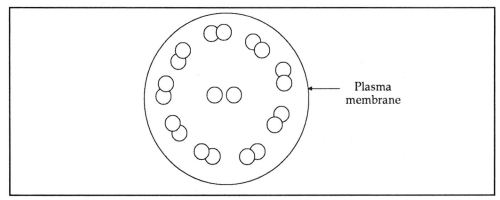

Plasma membrane

Figure 4.5 Diagram of a cross section of a cilium or a flagellum, showing two central tubules surrounded by nine outer tubules.

Π Compare these structures with those of the flagella of prokaryotes discussed in Chapter 1. We suggest you make a list. Consider their size and their chemical composition and the arrangement of these molecules. It will provide a useful form of revision.

You will notice that the two are quite different in terms of size and internal organisation. They also differ quite markedly in terms of the ways they are attached to the cell.

basal body

centriole

Inside the cell, cilia and flagella are attached to a basal body which has a structure identical as far as we can tell, to that of the centriole. (Compare with the hook and basal body of prokaryotes). Further, the basal body and centriole appear to be interconvertible in many instances. For example, *Chlamydomonas* has two flagella, each with a basal body. In preparation for division the flagella are resorbed into the body of the cell and the basal bodies become centrioles and migrate to a position near the nucleus where they generate the spindle microtubules. After mitosis, the centrioles migrate back to the plasma membrane, become basal bodies and regenerate the microtubules which form the flagellum. Even more striking is the example of division in frogs eggs, which are very large and, therefore, amenable to experimentation. Here the egg does not possess its own centriole but the sperm centriole enters during fertilisation and immediately begins to generate the mitotic spindle. Purified basal bodies from other species, when injected into frog eggs, behave in the same way. This is, therefore, powerful evidence for the centriole being the microtubule organising centre. We will discuss the role of the centriole in cell division in more detail in Chapter 8.

However, not all animals cells possess a detectable centriole and they do not occur at all in the cells of higher plants. Furthermore, whereas the nine triplet microtubules of a basal body give rise to the nine doublet microtubules of the cilium or flagellum this is not the case with the microtubules in the centriole. Close examination reveals that microtubules do not emanate from the centriole itself but from a densely staining **pericentriolar** material in which it is embedded, the pericentriolar material. Zones of similar material **material** appear to be the initiating region in those cells which lack a centriole. Perhaps this is the true organising centre for microtubules. One final point needs to be made. The microtubules in cilia and flagella appear to be more stable than normal cellular microtubules in that treatment with colchicine does not cause their dissolution. This and other evidence suggests that microtubules can be separated into two broad divisions based on their permanence. Clearly only those showing rapid production and dissolution will be affected by colchicine

---

SAQ 4.2

Fibroblast cells in culture were treated with colchicine for 2 hours, rinsed and treated with fluorescence-labelled anti-tubulin antibodies after three time intervals. They were examined using the fluorescence microscope and the following results were obtained (Figure 4.6). Note that anti-tubulin antibodies bind with tubulin.

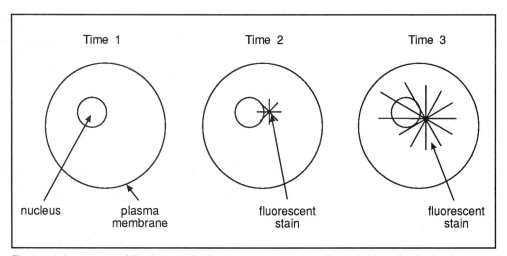

Figure 4.6 Appearance of fibroblasts using fluorescence microscopy. Time 1 is immediately after colchicine has been removed, Time 2 is 1 hour after colchicine removal; Time 3 is 2 hours after colchicine removal

What interpretation can you put on this experiment? Suggest ways to test your interpretations.

4.5 The structure of actin filaments

You are reminded that actin filaments are much smaller than microtubules.

G actin

The basic unit of an actin filament is a single globular protein, G actin, which has a molecular mass of 42000 daltons. Each molecule of G actin has a single Ca^{++} ion bound to it, which stabilises it, and one molecule of ATP. Polymerisation of G actin occurs to give F (filamentous) actin. ATP hydrolysis occurs during this polymerisation but energy is not required, as demonstrated by the ability of non-energy-releasing analogues of ATP to cause polymerisation. The rate of polymerisation is however much greater in the presence of ATP.

The polymerised filaments consist of two strands of G actin molecules twisted around each other forming a helix with 13.5 molecules per turn (Figure 4.7). These filaments can show a similar treadmilling to microtubules, with G actin being added to one end and removed from the other.

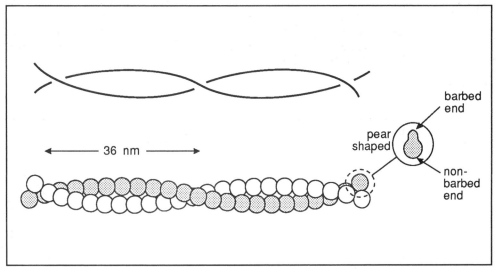

Figure 4.7 Drawing of an actin filament showing the helical arrangement of G actin monomers. G actin monomers are, in fact, pear shaped rather than spherical and this confers a structural polarity on the molecule which is confirmed by biochemical studies. The two ends are signified 'barbed' and 'non-barbed' as a result of interaction with myosin, another cellular protein.

acrosomal reaction

acrosomal process

Several pieces of evidence show that actin filaments can be assembled very quickly. In the acrosomal reaction of invertebrate sperm a thin tentacle called the acrosomal process is extended to penetrate the covering of the egg. The extension of this process is caused by the rapid polymerisation of G actin molecules to form a bundle of 25 actin filaments. The acrosomal process extends to greater than 50μm at a rate of 10μm per second.

| SAQ 4.3 |

To further impress upon you the amazing nature of this phenomenon, calculate how many A actin monomers are needed to produce an acrosome process of 48μm length assuming each actin monomer is 6nm long. For convenience we will ignore the effect of the helix.

microvilli

This rate of polymerisation is somewhat exceptional especially when you remember that the reaction of each monomer requires a molecule of ATP. A more normal rate of polymerisation (and depolymerisation) is shown by the microvilli, 1μm long finger-like extensions of the plasma membrane which occur on the surface of many cells eg those which line the small intestine (Figure 4.8).

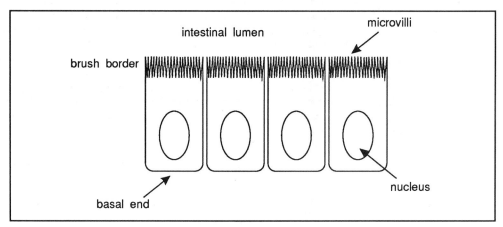

Figure 4.8 Microvilli on the cells of the small intestine. Each cell has several thousand microvilli which together are called the brush border. The area of the luminal surface is increased approximately 25x by the microvilli.

The actin filaments in the microvilli are held together in a tight bundle by accessory protein molecules, and the filaments at the tip of the microvillus are embedded in a zone of amorphous material which attaches the filaments to the plasma membrane. Further attachments between the plasma membrane and the filaments occur along the length of each microvillus. Thus the actin filaments form a well defined cytoskeleton.

stress fibres

adhesion plaques

Other bundles of actin filaments occur with less obvious function. Stress fibres are an example. These are bundles of actin filaments held together as in microvilli but present in the cytosol of cultured fibroblasts. These fibres terminate at specialised regions of the plasma membrane called adhesion plaques. Thus stress fibres and filaments in microvilli appear to be anchored to the plasma membrane, an obvious necessity if they are to exert any mechanical force. This function is much clearer in filaments in microvilli than in the stress fibres.

There is also evidence that the ordinary actin filaments as opposed to the stress fibres, are attached to the plasma membrane but nothing is known about the relationship between ordinary and stress fibres.

4.6 Drugs also modify actin polymerisation

The occurrence of changes in actin polymerisation are easy to detect in situations where the changes are large and relatively permanent (eg the acrosome reaction) but not in more normal situations. As with the study of microtubules, use is made of drugs. The cytochalasins are a group of compounds, excreted by various fungi, which bind to the barbed end of actin filaments. Actin filaments show the treadmilling phenomenon shown by microtubules and grow most actively at the barbed end. The binding of cytochalasin stops polymerisation. In contrast phalloidin, which is a product of another fungus, the poisonous toadstool *Amanita phalloides*, prevents depolymerisation by binding to them along their length. These drugs are, therefore, tools which can be used in the diagnosis of processes involving the formation and deformation of actin filaments.

cytochalasins

phalloidin

Amanita phalloides

| SAQ 4.4 | Under which of the following conditions do you think actin filaments might best be visualised. Give your reasons.

1) Treatment with fluorescence-labelled anti-actin antibodies followed by fluorescence microscopy.

2) Treatment with phalloidin followed by treatment with fluorescence-labelled anti-phalloidin antibodies followed by fluorescence microscopy.

4.7 Intermediate filaments

These are tough and durable fibres which appear to be straight or gently curving in electron micrographs. They are found especially in places exposed to mechanical stress eg in muscle cells and the axons of nerve cells. They are very stable, in contrast to microtubules and actin filaments, and appear to require enzyme action for their degradation. When cells are extracted with non-ionic detergents intermediate filaments are the only fibres not dissolved. The name cytoskeleton was originally invented for this group of insoluble fibres. Biochemically they are a diverse group containing fibres whose monomers vary from 40 000 to 200 000 daltons. They have in common the fact that their basic structure is not a globular protein but has a more thread-like appearance. They associate side by side in trimers in a rope-like manner similar to the very stable connective tissue fibrous protein collagen (see Chapter 10). This type of arrangement would produce fibres with high tensile strength (Figure 4.9).

stability of intermediate filaments

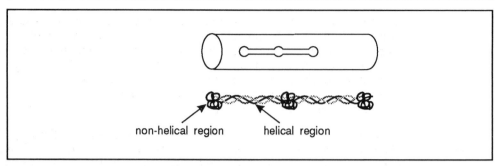

Figure 4.9 Model of structure of an intermediate filament. Helical regions are responsible for the lateral binding which forms the fibre. Non-helical regions may be involved in cross linking with other filaments.

∏   What do you think is the function of the intermediate filaments? Is there a separate role for the non-helical portions? (Have a guess!).

Progress here is limited by the lack of suitable reagents which act as colchicine and cytochalasin do with microtubules and actin filaments respectively; therefore it is not yet possible to selectively remove intermediate filaments from a cell. It is tempting to speculate however that they are involved in withstanding tension and that their non-helical zones allow cross reaction with other filaments both intermediate and others. The crosslinking role in microtubules and actin filaments is taken by specific accessory proteins. If the non-helical zones behave as suggested it would appear that intermediate filaments have in-built accessory proteins.

4.8 Organisation of the cytoskeleton

So far in this chapter we have treated the three groups of filaments as if they were entirely separate from each other. It is obvious, however, that the different parts of the cytoskeleton must be linked together and their functions coordinated. Unfortunately relatively little is known about links and interactions between the three groups and conventional thin section electron microscopy will not help.

Thicker sections of cells are difficult to interpret but it does seem that actin and intermediary filaments are heavily intertwined and cross linked. Microtubules are also thought to be connected to the actin/intermediary filament network.

4.9 Do microtubules have an organising role?

Certain evidence suggests a controlling role for microtubules in the organisation of the cytoskeleton. They are commonly found aligned parallel with the long axis of cells such as nerve axons and it has been demonstrated in some case that their presence is necessary for the maintenance of polarity. Thus application of colchicine to growing nerve axons stops microtubule extension and axon growth. During the development of the endothelial lining of the small intestine (Figure 4.7), cells grow in the basal to apical direction (ie they have a polarity) and the application of colchicine prevents this growth.

Microtubules also seem to influence the distribution of intermediate filaments. In cultured fibroblasts intermediate filaments spread out in a radial direction from a region near the nucleus, a distribution which is very similar to that of microtubules. Within 10 minutes of the application of colchicine, the microtubules have disappeared and over the next few hours the intermediary filaments retract to form a dense mass near to the nucleus. When the colchicine is washed away the microtubules rapidly reappear and the intermediary filaments slowly return to their original distribution.

contractile ring

Some evidence also links microtubules with actin filaments. During cell division (see Chapter 8) a contractile ring forms which cleaves the cell into two. This ring consists of actin plus myosin filaments and its location in the cell appears to be under the control of microtubules.

membrane ruffling

Finally, in cells in culture, the phenomenon of membrane ruffling, a sweeping wavelike motion, often appears to occur in regions opposite to the end containing microtubules. The ruffling membrane occurs at the forward edge of a migrating cell and actin filaments are implicated in this cell migration from the fact that cytochalasin prevents it. If microtubules are disrupted by colchicine, membrane ruffling occurs in an apparently random fashion and the cell now wanders as if lost suggesting that the microtubules are in some way involved with the directional operation of the actin filaments. This evidence might point to a controlling role for microtubules but it must be remembered that actin filaments are the only ones for which there is good evidence of plasma membrane anchorage, an obvious prerequisite, as already mentioned.

Thus we might conclude by saying that the cytoskeleton consists of a network made up of three types of filament, stabilised by attachment of the actin filaments to the plasma membrane. In spite of this very important role for actin filaments microtubules appear to be the overall controlling influence. However we noted earlier that microtubules themselves are organised by the centriole, which perhaps should be considered to be the headquarters for these aspects of cell behaviour. The centriole is sometimes referred to as the cell centre. In support of this it has been observed that in migrating cells, the centriole is on the side of the nucleus closest to and facing the advancing ruffling membrane and that one of the centriole half pairs is pointing in the direction of the moving membrane.

SAQ 4.5		

Insert a cross where the statement applies.

Statement	Actin filament	Microtubule	Intermediate filament
1) Is made up of globular protein			
2) Is a fibrous protein			
3) Capable of self assembly and disassembly			
4) ATP is 'incorporated' into structure			
5) GTP is 'incorporated' into structure			
6) Form of polymer			
a) 2 stranded helix			
b) hollow tube			
7) Diameter of filament			
a) 25nm			
b) 8nm			
c) 10nm			
8) Requires accessory proteins to make lateral connection			
9) Monomer of more than one size			
10) Is attached to the plasma membrane			
11) Involved in brush border structure			
12) Involved in the structure of cilia and flagella			

4.10 Plant cell walls

As noted in section 4.1, the shape of plant cells is governed by their cell walls. We will now describe their structure and attempt to explain how their shape is generated.

primary and secondary wall

Three regions in a plant cell wall can be recognised; the middle lamella, the primary wall and the secondary wall (Figure 4.10).

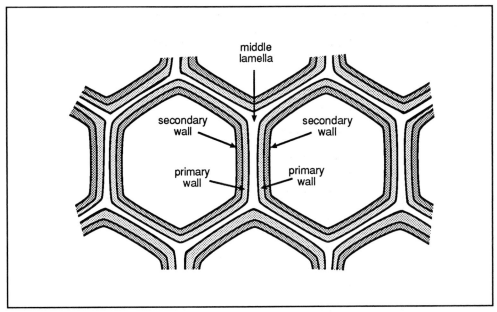

Figure 4.10 Diagram of plant cells to show the common middle lamella and the primary and secondary walls. Note: cell contents omitted and wall thickness increased for clarity.

The middle lamella is a joint wall between adjacent cells. The next layer laid down is the primary wall and this and the middle lamella are the only ones present in cells as they grow. The secondary wall is usually thicker than the middle lamella and primary wall together and its deposition stops the growth of the cell.

SAQ 4.6

Regarding cell wall structure of plants, which of the statements below apply to 1) the middle lamella, 2) the primary wall, 3) the secondary wall?

a) stops cells from growing;

b) holds adjacent cells together;

c) is present in growing cells;

d) is the thickest part of the wall;

e) is present throughout the whole wall;

f) is the last part of the wall to be laid down.

cellulose

hemicellulose

pectin

glycoprotein

The basic structure of the wall is that of long tough fibres, approximately 50nm in diameter and several μm long, embedded in a matrix of polysaccharide and protein; similar in principle to that of resin-impregnated fibre glass or steel-reinforced concrete. The long fibres are made of cellulose and the matrix consists of pectin, hemicellulose and wall glycoprotein. Although the basic structure is simple, there is considerable complexity in many of the components and we are still not sure about the exact nature of the crosslinks between them.

4.11 What is the structure of the fibre?

β-1,4 links The long fibres of the wall are cellulose which consists virtually 100% of glucose molecules joined end to end by β-1,4 glycosidic links (Figure 4.11). The orientation of the link at the 4 carbon position is slightly twisted and so is the link at the 1 carbon. If the molecules were formed by α-1,4 links these twists would accentuate each other and would result in a gentle helix, as shown by starch in a starch grain. The β orientation at the 1 carbon position twists the other way, counteracting the twist on carbon 4. This brings the molecule back to the straight line. Thus the basic unit of cellulose is a polymer of glucose which forms long straight chains. Many hundreds of these associate with each other by hydrogen bonds forming the very strong cellulose fibre. Polymers of glucose linked by α-1,4 links (eg starch and glycogen) are used as storage forms of carbohydrate.

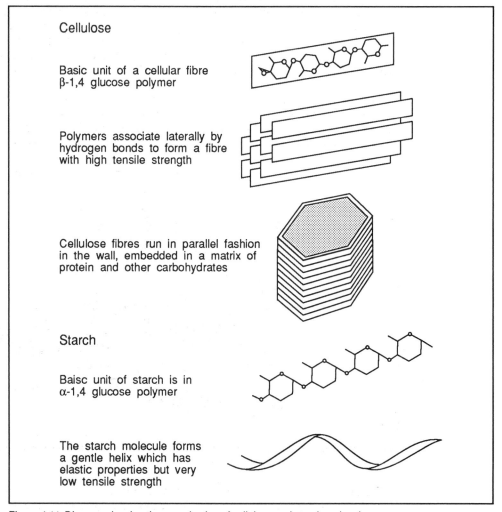

Cellulose

Basic unit of a cellular fibre
β-1,4 glucose polymer

Polymers associate laterally by
hydrogen bonds to form a fibre
with high tensile strength

Cellulose fibres run in parallel fashion
in the wall, embedded in a matrix of
protein and other carbohydrates

Starch

Baisc unit of starch is in
α-1,4 glucose polymer

The starch molecule forms
a gentle helix which has
elastic properties but very
low tensile strength

Figure 4.11 Diagram showing the organization of cellulose and starch molecules.

SAQ 4.7	Glucose can form polymers by forming links between its carbons at position 1 of one molecule and the carbon at position 4 on another. The orientation of the link at carbon 1 can be α or β.

1) What is the significance of the β-1,4 link in cellulose?

2) To what use are α-1,4 linked polysaccharides put?

4.12 How are polymers arranged in the wall?

hemicelluloses

pectin

Hemicelluloses are a diverse group of branched polysaccharides made up of several different monomers. They coat the cellulose fibres and bind to them by hydrogen bonds. Pectins are the third group of polysaccharides. These too are somewhat heterogeneous, with short side chains but they have in common a considerable number of galacturonic acid residues in their backbone, from which the carboxyl groups protrude laterally. These can readily form cross bridges with divalent cations, particularly calcium, and are transformed from a liquid into a gel. This material is often used in jam making. The reader is referred to the BIOTOL text 'The Fabric of Cells' for greater detail at the chemistry of these cell wall constituents.

The pectin polymers are covalently attached to hemicellulose molecules and also to the wall glycoprotein molecules, which themselves can form cross bridges with other pectin molecules. The result is a complex network of interlinked polymers (Figure 4.12).

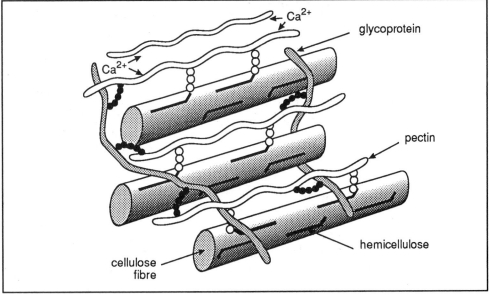

Figure 4.12 Model of primary cell wall structure to describe possible arrangement of cellulose fibres and the matrix of hemicellulose, pectin and glycoprotein.

The structure of cellulose confers great strength in the axial direction, and the orientation of the cellulose fibres in the wall of the growing cell is the primary factor controlling the shape of that cell. Expansion can occur only at right angles to the cellulose fibres. It is not surprising, therefore to find that the cellulose fibres in the primary wall are predominantly perpendicular to the cell's long axis. Growth is considered to involve the breaking of the links between the pectin and the hemicellulose which allows the sliding apart of the cellulose fibres.

<table>
<tr><td>SAQ 4.8</td><td>

Regarding plant cell wall structure, which of the statements below apply to 1) cellulose, 2) hemicellulose, 3) pectin and 4) wall glycoprotein?

a) made of α-1,4 linked glucose residues;

b) part of the matrix of the wall;

c) linked to other components of the wall by hydrogen bonds;

d) can form cross links with Ca^{2+} ions;

e) hydrogen bonds are very important in its structure;

f) major factor in governing cell shape;

g) breakage of links between hemicellulose molecules and this component is involved with cell growth;

h) directly attached to only one other wall component.

</td></tr>
</table>

lignin

The secondary wall differs from the primary wall in containing more cellulose and less pectin and glycoprotein. The orientation of cellulose fibres here is in no particular direction. Additional material may be laid down in the wall (eg lignin; composed of phenolic components) which strengthens the matrix of the whole wall. Wood consists of lignified plant cells and it is this binding together of the cellulose fibres with lignin that confers strength to the material.

4.13 How is the wall laid down?

matrix polymers

cellulose microfibrils

It is not surprising to find that the fibres of the wall are laid down by a method quite separate from the matrix materials. Evidence involving the exposure of growing cells to radioactive glucose shows that some of this material is transported to the Golgi apparatus, converted into precursors of the matrix polymers, packaged into Golgi vesicles, released to the wall area by exocytosis and is subsequently polymerised there. There is no spatial organisation in this activity, it occurs all over the surface of the growing cell. Cellulose fibres are laid down in a particular orientation in primary walls, namely perpendicular to the axis which becomes the long axis of the cell. The synthesis of the long poly-glucose chains is brought about by tightly associated groups of plasma membrane proteins. The poly-glucose chains are released into the wall compartment where they bind laterally by hydrogen bonds to form the thick fibres. The proposed mechanism for this is fascinating. The release of the poly-glucose chains apparently pushes the 'synthetic station' along in the fluid membrane and microtubules appear to be involved in the control of direction of this migrating unit (Figure 4.13). In normal tissues, microtubules are found running close to the plasma membrane perpendicular to the long axis of the cell. Thus microtubules and cellulose microfibrils are arranged in parallel (Figure 4.13). Further evidence for a causative link here comes from tissues treated with ethylene, a plant growth hormone. This alters the direction of growth of cells and causes them to grow short and wide rather than long and thin. In such tissues, the cellulose microfibrils and the microtubules are found to have changed their orientation by 90°. Thus it appears that microtubules are acting as tram lines along which cellulose is laid down. Unfortunately these microtubules appear to be of the stable kind and are not disassembled as a result of colchicine treatment, thus their controlling influence has not been absolutely confirmed.

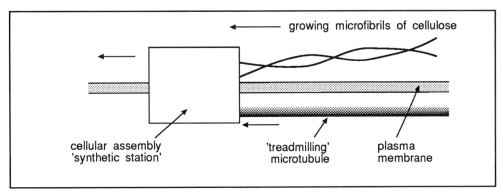

Figure 4.13 Diagram showing the growth of cellulose being directed by microtubules.

An interesting point arises from the study of pollen tubes which grow by tip growth. Here the cellulose fibres are randomly arranged in the wall and growth appears to be governed by the rate of incorporation of material from Golgi vesicles. Treatment with cytochalasin, but not colchicine, inhibits pollen tube growth suggesting the involvement of actin filaments. In the presence of cytochalasin, Golgi vesicles are retained close to the Golgi apparatus and this raises the possibility that actin filaments are involved in vesicle transport to the plasma membrane. If actin filaments are involved in the deposition of matrix materials in pollen tubes, they might also be involved in this process in other cells.

4.14 Concluding remarks

Most of the work studying the cytoskeleton has been carried out on animal cells where microtubules, actin filaments and intermediate filaments appear to be jointly responsible for the control of cell shape. Plant cell shape is governed by the shape of its cell walls but here to there is evidence that this is also controlled by microtubules and actin filaments. To this extent cell shape in plants and animal cells is governed by the same principal cytoplasmic components.

Summary and objectives

In this chapter we have discussed the factors which govern cell shape. Although many differences between plant and animal cells were described, some underlying principles were established. Central to the control of cell structure, were the roles of microtubules and actin filaments. The use of drugs to analyse the involvement of these two components in a variety of cell activities was also described. Details of the structure of plant cell walls were also covered.

Now that you have completed this chapter, you should be able to:

- describe the meaning of the term cytoskeleton and show a knowledge of the structure of microtubules, actin filaments and intermediate filaments;

- show evidence of understanding the role and properties of the cytoskeletal components by interpreting the results of experiments;

- demonstrate a knowledge of the role of the middle lamella and the primary and secondary walls in plant cell wall structure;

- show an understanding of the significance of the nature of biochemical links in the properties of glucose polymers.

Chloroplasts and Mitochondria

Chloroplasts and Mitochondria

5.1 Introduction

All organisms need a supply of energy to drive the biochemical and physical processes which go on within them and which are essential to their survival. Energy is contained within all large molecules, particularly in fats and carbohydrates, but this energy is not available for cellular work until the compounds have been degraded releasing the cellular energy currency ATP. Chloroplasts and mitochondria are the two organelles involved with energy metabolism. They are both involved in ATP production and elegant research over the passed 20 years has shown that the mechanism by which this

ATP generation   is achieved is the similar in the two organelles. Further, prokaryotes need to be able to generate ATP and they may use a similar mechanism (given the name chemiosmosis).

chemiosmosis   An exception is found in some prokaryotes which grow in anaerobic conditions. We will not be discussing these further in this text. The reader is referred to 'Principles of Cell Energetics' in the BIOTOL series for a fuller discussion of the generation of ATP in these systems.

Chemiosmosis is a very powerful unifying principle in biology. Basically the process begins with high energy electrons, either formed by excitation with sunlight or captured within high energy food compounds. These electrons are passed along a chain of

electron   electron carriers (electron transport chain), which are held within membranes. As the
transport chain   electrons pass along the chain their 'energy level' falls in a step-wise manner and the energy released moves protons across the membrane. This generates an

electrochemical   electrochemical (proton) gradient which can either be used to drive local active
gradient   transport systems or to generate ATP which can be utilised in sites remote from its production. This process has been stylised in Figure 5.1. Chloroplasts and mitochondria use most of the energy for the generation of ATP.

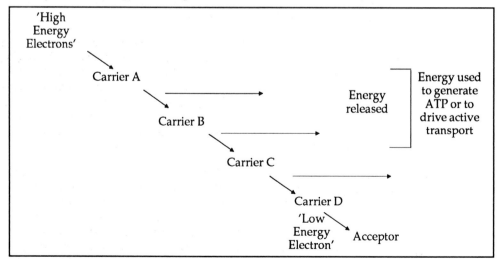

Figure 5.1 Stylised representation of ATP generation coupled to an electron transport chain.

Chloroplasts and mitochondria each contain DNA which is consistent with modern ideas of their origin. It is suggested that the two organelles have evolved from prokaryotes that were engulfed by primitive eukaryotes and somehow managed to avoid being digested by lysosomes. This resulted in the development of a symbiotic relationship (ie one beneficial to both partners, ensuring their survival) between the host cell and its incumbent prokaryote. There is considerable evidence to support this hypothesis which we will examine at the end of the chapter. We will however, first concentrate on the structure and function of these two types of organelles.

5.2 Organisms can be classified according to their mode of nutrition

autotroph

phototroph

chemotroph

heterotroph

Organisms which are able to synthesise all the organic compounds they need simply by using inorganic compounds in their environment are called autotrophic organisms. They may use light energy to drive these processes in which case they are called phototrophs or they may be able to use energy derived from the oxidation of inorganic compounds, in which case they are chemotrophs. Organisms which depend on a supply of already-formed organic compounds are called heterotrophs. Many bacteria are chemotrophic; plants are phototrophic and animals heterotrophic. We will concentrate on the last two and will begin with phototrophs which produce organic compounds in the process of photosynthesis.

5.3 Where does photosynthesis take place?

In eukaryotes, photosynthesis takes place in chloroplasts. In prokaryotes, a variety of other structures are employed. Here, we will confine ourselves to the description of the photosynthetic apparatus of higher plants. Although numbers vary, a typical palisade cell (leaf cell) may contain about 40 chloroplasts. They are usually 3-10μm long and about 2-4μm across and consist of about 50% dry weight of protein, 40% lipid and a variety of smaller water soluble molecules. We shall learn that the lipid fraction contains many pigments such as carotenoids and chlorophylls as well as the major lipids of membranes (phospholipid, galactosyl diglycerides and sterols).

thyllakoids

grana

stroma

The outer limit of each chloroplast is defined by a double membrane envelope. Within this envelope is a large number of membranes or lamellae (thyllakoids). These lamellae may be closely stacked into grana. These grana are interconnected by extensions of the lamellae (Figure 5.2). The grana are the sites of oxygen evolution and of ATP synthesis. The thyllakoids separate the chloroplast into two separate components - the thyllakoid compartments and the stroma. We find the enzymes responsible for CO_2 fixation in the stroma. The stroma also contains ribosomes, nucleic acids and the enzymes involved in fatty acid synthesis.

C_4 plants

There are some variations to this basic structure. The most common one, found in so called C_4 plants, is for the lamellae to extend for the whole length of the organelle. C_4 plants are so named because of the metabolic pathway used for CO_2 fixation which is characteristic of many monocotyledons including the tropical grasses, sugar cane and corn. These variations are quite important in terms of crop productivity and in the details of the biochemical reaction involved, but are of secondary importance to the basic principles of photosynthesis. Here we will confine ourselves to these basic principles.

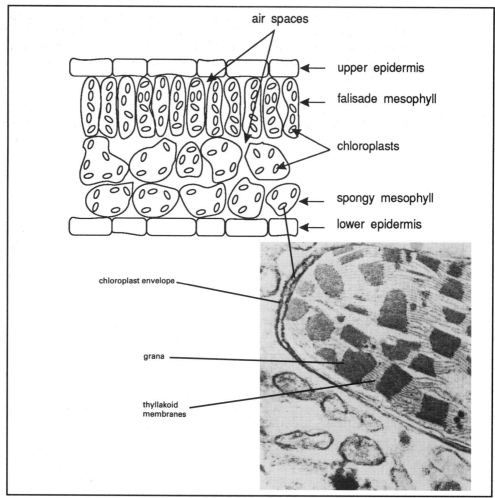

Figure 5.2 a) Diagram of a cross section of a leaf showing extensive air spaces around the cells and numerous chloroplasts. b) High magnification of part of a chloroplast showing the chloroplast envelope and the internal membrane system forming thyllakoids and grana (x 20 000).

5.3.1 The overall reactions of photosynthesis

An overall equation for the photosynthesis of higher plants can be written:

$$6 CO_2 + 12 H_2O \rightarrow C_6H_{12}O_6 + 6 O_2 + 6 H_2O$$

In this process carbon dioxide is reduced using hydrogen atoms from water producing carbohydrate along with the release of O_2. 12 molecules of H_2O are included because all of the O_2 released comes from water. The process can be measured by the rate of absorption of CO_2 or the rate of release of O_2. We can call such photosynthesis oxygenic as molecular oxygen is generated. Other types of photosynthesis are known amongst prokaryotic forms.

oxygenic

light and dark reactions

Writing a single equation for photosynthesis hides the fact that the process can be divided into two parts; the light-dependent part and the light-independent part. The latter is sometimes called the dark reaction. This is acceptable as long as you realise that the dark reaction in not dark-requiring.

In the light reaction the molecule of water is split, using light energy captured by a chlorophyll molecule. O_2, ATP and reduced cofactors, in the form of NADPH, are formed. In the dark reaction, the NADPH and ATP are used to reduce CO_2 to produce a sugar, an energy-rich compound. It is important to realise that the NADPH and ATP used have come only from the light reaction.

SAQ 5.1

Below is an incomplete diagram describing the inputs and outputs of photosynthesis and the relation between the light and dark reactions. Fill in the boxes labelled A-F. Use the following labels ATP, NADPH, dark reaction, O_2, sugar, light reaction

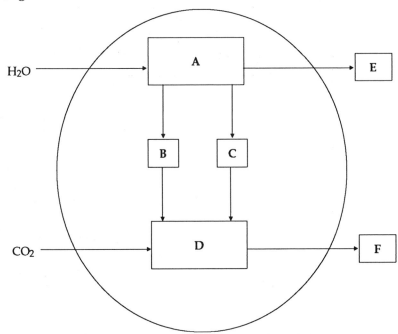

5.3.2 Absorption spectra and action spectra help our understanding of photosynthesis

absorption spectrum

action spectrum

If a tissue absorbs light we can find out its absorption spectrum, which shows the degree to which light of different wavelengths is absorbed. If a process is catalysed by light we can find out its action spectrum, which is the effect of different wavelengths of light on the rate of the process. Figure 5.3 shows the action spectrum of the process of photosynthesis and the absorption spectrum of the major chloroplast pigments, chlorophylls a) and b) and carotenoids.

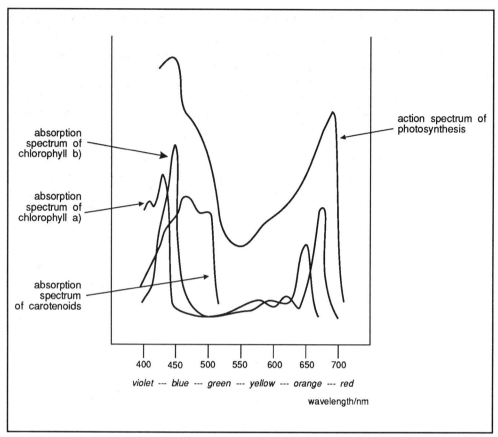

Figure 5.3 Absorption spectrum of carotenoids, chlorophylls a) and b) and the action spectrum for photosynthesis.

SAQ 5.2

1) What does Figure 5.3 show us about photosynthesis?

2) Why is a leaf green?

5.3.3 Fractionation studies allow the location of the light and dark reactions to be decided

If a tissue is carefully homogenised, intact chloroplasts can be isolated from the homogenate by density gradient centrifugation. If these chloroplasts are exposed to a solution of higher water concentration than that of the stroma, water will enter by osmosis and the chloroplasts will swell and eventually burst. It is possible to separate the lamellae and grana membranes from the chloroplast envelope by centrifugation and study them separately from the stoma. Such studies show that the components of the light reaction are found on the membranes whereas those of the dark reaction are present in the stroma. This approach has allowed the two processes to be studied in isolation.

light reactions on membranes

dark reactions in the stroma

5.3.4 The light reaction contains a cyclic and a non-cyclic pathway

cyclic and
non-cyclic
photo-
phosphorylation

It was stated above that the products of the light reaction were O_2, ATP and NADPH. Detailed studies of the light reaction reveals a second type of light reaction where the product is ATP alone. This latter mechanism operates in the form of a cycle and is called cyclic photophosphorylation. The original (O_2 generating) reaction does not operate as a cycle and is called non-cyclic photophosphorylation. The term photophosphorylation refers to the addition of inorganic phosphate to ADP forming ATP, using light energy.

5.3.5 Non-cyclic and cyclic photophosphorylation

Figure 5.4 shows a scheme for non-cyclic photophosphorylation. Do not worry for the moment about the redox potential scale on the left; we will refer to this later.

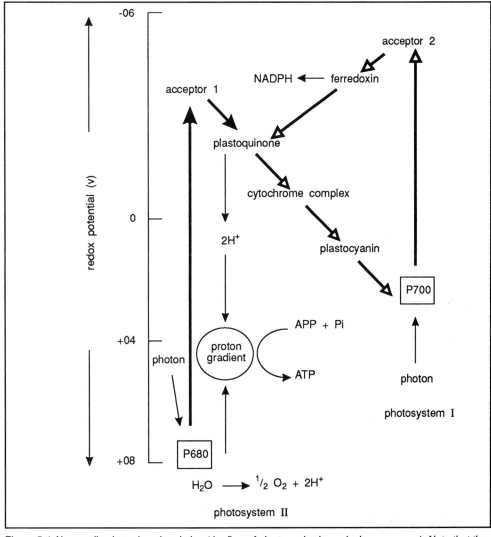

Figure 5.4 Non-cyclic photophosphorylation (the flow of electrons is shown by heavy arrows). Note that the splitting of water and the transfer of electrons from plastoquinone to the cytochrome complex generate protons. These are released on one side of the photosynthetic membrane, thus producing a proton gradient. This gradient is used to drive the synthesis of ATP from ADP and inorganic phosphate.

Let us examine non-cyclic phosphorylation and its associated electron transport in more detail. As you read the text refer to Figure 5.4. All of the chlorophyll a and b molecules are bound to protein molecules and one of the chlorophyll a/protein complexes is called P680, referring to the wavelength, it maximally absorbs light. This complex absorbs a photon of light which activates an electron which is used to reduce acceptor 1, leaving the chlorophyll molecule (temporarily) with a net positive charge. The P680 complex is sometimes referred to as photosystem II. At the same time a second chlorophyll a/protein complex, P700, absorbs a photon and the activated electron reduces acceptor 2, leaving P700 with a net positive charge. The P700 complex is also referred to as photosystem I. Acceptor 1 now transfers its electron through an electron transport chain consisting of plastoquinone, a cytochrome complex and plastocyanin and neutralises the positive charge on P700. We will not concern ourselves with the details of the molecular structure of these electron transport components here except to make the following points. Plastoquinone can be reduced to plastoquinol thus:

P680

P700

plastoquinone

cytochrome

Cytochromes are proteins which contain tetrapyrolle groups as functional groups. Their reduction can be represented by the reaction:

or more simply:

$$Fe^{3+} + e^{-1} \rightleftarrows Fe^{2+}$$

Note that the figures showing the tetrapyrolle groups have been simplified.

plastocyanin

Plastocyanin is a copper containing protein which can be oxidised and reduced. We can represent this reaction by:

$$plastocyanin\ (oxidised) + e^{-1} \rightarrow plastocyanin\ (reduced)$$

ferredoxin

The electron on acceptor 2 is transferred through an intermediary, ferredoxin; to $NADP^+$ to form NADPH. Ferredoxin is a protein which is rich in iron and sulphydryl groups. We can represent its oxidation and reduction by:

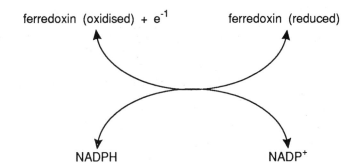

Finally the positive charge on P680 is neutralised with an electron released during the oxidation of water, a process coupled with O_2 release. We can write this reaction as:

$$H_2O \rightarrow 2e^- + 2H^+ + O_2$$

If we examine the operation of the electron transport chain closely we will see that when plastoquinone is reduced, hydrogen ions and electrons are required but, when it is oxidised, the hydrogen ions (protons) are released because the subsequent components only accept electrons. In a reaction which we will examine later, the protons released here and in the breakdown of water are used to drive the manufacture of ATP in the process of chemiosmosis. We can see, therefore, that the products of the non-cyclic scheme are NADPH, O_2 and ATP.

In cyclic photophosphorylation the electrons transferred from acceptor 2 to ferredoxin are not passed on to NADPH but to plastoquinone. From there they move along the electron transport chain back to the vacant space in P700 (follow this on Figure 5.4 using the unfilled arrows). Only photosystem 1 is involved in this process. No NADPH or O_2 production occurs. The protons released from the oxidation of plastoquinol, however, are used to drive the synthesis of ATP. This is the only product of cyclic photophosphorylation. The chloroplast therefore posesses an alternative mecahnism to generate extra ATP.

We do not know how the cell controls which of the two schemes it will use but it is important for later purposes to realise that, by suitable use of the two, various proportions of ATP to NADPH can be produced.

5.3.6 The significance of the redox potential

electron affinities

reducing power

Compounds which accept and transfer electrons are very common in biology, especially in energy metabolism. These compounds differ from each other in the ease or readiness with which they accept electrons. We say that they have different electron affinities. The scale of electron affinity is called the redox potential scale, measured in volts, and the compounds which occur in biology have values which range from -0.6 to +0.8 volts. The higher the redox potential the higher the electron affinity. Putting this another way, the lower the redox potential the greater is the tendency to donate electrons ie the greater is the reducing power. This term is quite meaningful because the energy level of an electron is lower as the redox potential is raised. In proposing any scheme involving the transfer of electrons the redox potential of donors and recipients must be 'correct' or the scheme will not be feasible. Figure 5.4 has a scale of redox potential. Use the information there to, answer SAQ 5.3.

SAQ 5.3	Which of the following are energetically feasible?

Transfer of an electron from:

1) acceptor 1 to acceptor 2;

2) acceptor 2 to acceptor 1;

3) plastocyanin to plastoquinone;

4) NADPH to plastoquinone;

5) ferredoxin to plastoquinone;

6) water to $NADP^+$.

use of inhibitors It is important to demonstrate that any proposed electron transport chain shows the correct sequence of redox potentials. You must however realise, that even if this is so, it does not prove that the chain operates as you may have proposed. This needs careful work and follows two broad methods. Firstly inhibitors are often available which specifically bind to a component and prevent its oxidation and reduction. If this component is part of electron transport chain, the inhibitor should block the pathway. Use can also be made of the fact that the wavelength of maximum absorption of light differs in some compounds between the oxidised and reduced forms. If the component is part of an electron transport chain it is sometimes possible to detect this by looking for absorption changes.

5.3.7 Are chlorophyll molecules loners or groupies?

You may have the impression from the above that chlorophyll molecules act independently of one another. This is not true. They are grouped into clusters of several hundred molecules which co-operate with each other in the light-harvesting process. **antenna complex** These clusters are usually called the antenna complexes. All chlorophyll molecules occur as chlorophyll/protein complexes and several different proteins occur in such complexes. Combination of the chlorophyll molecule with its protein is very important because it slightly modifies the wavelength of maximum light absorption. Thus these antenna complexes, by virtue of possessing several different chlorophyll/protein combinations, can absorb light over a much wider range of wavelengths than would be the case with just pure chlorophyll a and b. This broadening of light absorption is further extended in many plants by the inclusion of other pigments (eg carotenoids). Each antenna complex has a single P680 or P700 reaction centre and approximately 250 chlorophyll molecules. The various components are so arranged that the energy of any captured photons can be transferred from chlorophyll to chlorophyll and finally to the reaction centre. Each absorbed photon activates an electron on a pigment/protein **reaction centre** complex and this causes activation of an adjacent complex which passes the energy on until eventually it reaches the reaction centre. We can represent this flow thus:

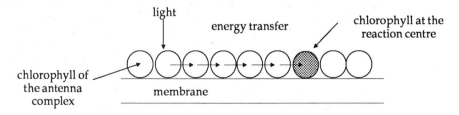

5.3.8 A proton gradient is formed in the light reaction

importance of
intact
thyllakoids

As we have seen, it is possible to break open chloroplasts, separate the thyllakoids from the stroma and show that the light reaction is catalysed by components which are bound into the thyllakoid membranes. Although broken thyllakoids can reduce $NADP^+$ and breakdown water releasing O_2, only in the presence of intact thyllakoids can ATP formation take place. This is because, during the light reaction, hydrogen ions are transported into the thyllakoid space where they increase in concentration to the point where the internal pH is about 5.0. The pH of the stroma is approximately 8.0 so the operation of the light reaction generates a gradient of about 3.0 pH units. A proposed scheme to explain how this is brought about is shown in Figure 5.5.

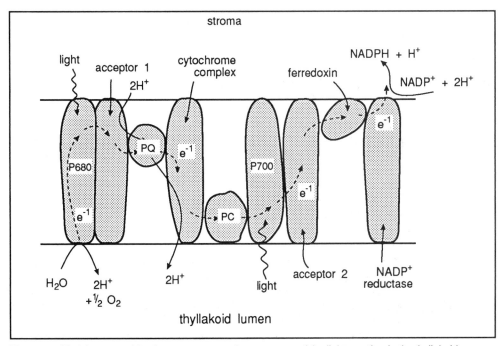

Figure 5.5 Model proposed for the arrangement of components of the light reaction in the thyllakoid membrane and for the pathway taken by electrons. The reaction centres P680 and P700 are the only parts of the antenna complexes shown. Phospholipids have been omitted for clarity. PQ = plastoquinone; PC = plastocyanin.

Protons are generated in two places in Figure 5.5: during the oxidation of water and during the reduction and oxidation of plastoquinone (PQ). Although we do not know the exact mechanism of water oxidation, we do know that it happens inside the thyllakoid, thus releasing H^+ ions into the thyllakoid lumen. The reduction of plastoquinone takes place on the thyllakoid lumen side. Since the reduction of plastoquinone to plastoquinol involves the uptake of H^+ ions while the oxidation of plastoquinol causes the release of H^+ ions, the oxidation/reduction of plastoquinone effectively pumps H^+ ions across the membrane. Plastoquinone is a small fat soluble molecule, however, and it does not protrude from the membrane on either side. Thus it cannot directly absorb protons from the stroma. This is an unresolved problem and one possibility being considered is that plastoquinone 'steals' protons from acceptor 1, which then makes up its loss by taking protons from the stroma. The reduction of the cytochrome complex involves only acceptance of electrons from reduced plastoquinone and does not require H^+. The H^+ ions are considered to be released into the thyllakoid space, the electrons being retained by the cytochrome.

5.3.9 ATP is generated by another membrane component

The final part of the light reaction concerns the use of the proton gradient to generate ATP. This involves a membrane-bound component which spans the membrane and a second component attached to it which protrudes into the stroma (Figure 5.6).

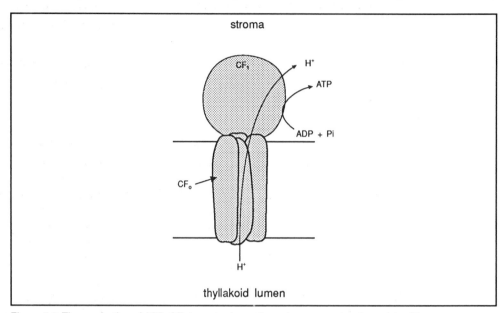

Figure 5.6 The production of ATP. CF_o is a membrane-bound component and consists of four polypeptides. CF_1 is made up of five polypeptides.

ATP synthetase

The CF_o/CF_1 complex shown in Figure 5.6 synthesises ATP and it is, therefore, referred to as ATP synthetase. Of course, it does not mind how the proton gradient is generated and thus it is contributed to by both non-cyclic and cyclic photophosphorylation systems. As you can imagine it is not difficult to physically knock CF_1 off the thyllakoid membrane. CF_o is involved in proton transfer across the membrane. CF_1 couples this with ATP synthesis.

∏   What effect do you think removal of CF_1 has on ATP synthesis?

It completely inhibits it and the protons simply leak out through CF_o.

SAQ 5.4

Can you think of ways of preventing ATP synthesis, other than by turning off the light?

SAQ 5.5	Match the appropriate process to the statements.

Process	Statements
1) cyclic photophosphorylation	a) forms NADPH, O_2 and ATP
2) non-cyclic photophosphorylation	b) involves only photosystem 1
	c) involves photosystems 1 and 2
	d) forms ATP only
	e) involves the oxidation of water
	f) involves P680

5.3.10 The study of the dark reaction involves very short experiments

We have seen that the dark reaction involves the conversion of CO_2 into sugars and this obviously involves some intricate reactions because of the differences in chemical complexity between the two. The details of this process were worked out by exposing plants to $^{14}CO_2$ and then isolating and identifying all the radioactive compounds. The plan was to arrange these into a reaction sequence which would describe the pathway of production of the sugar. However, it was discovered very quickly that so many compounds became labelled that it was impossible to sort them out. The strategy subsequently chosen was to use single-celled green algae as the photosynthesising plants and to expose them to light for just a few seconds. They were then dropped into boiling alcohol to stop the reaction. The algae were homogenised and the soluble compounds separated by chromatography and identified. These studies showed that the compound which became radioactively labelled first was a three-carbon compound, 3-phosphoglyceric acid.

∏ Bearing in mind that $^{14}CO_2$ had been applied to the cultures, what can you propose about the compound which reacted with CO_2 to form 3-phosphoglyceric acid?

Think of an answer before reading on. It is natural to suggest that a two-carbon acceptor reacted with CO_2 to form 3-phosphoglyceric acid. Unfortunately no such compound could be found.

∏ Can you suggest an alternative?

Calvin cycle

Again, try to think of an answer. An alternative is that a five-carbon compound reacted with CO_2 forming two molecules of 3-phosphoglyceric acid. This was, in fact, shown to be the case. Subsequent work led to the proposal of the Calvin cycle (named after the researcher who did much to elucidate the reactions) to describe the events of the dark reaction (Figure 5.7).

SAQ 5.6

Complete the following equation:

1) $3CO_2 + ?\,ATP + ?\,NADPH \rightarrow$ glyceraldehyde 3-phosphate $+ ?\,ADP + ?\,Pi + ?\,NADP^+$

2) How does the plant provide ATP and NADPH in unequal amounts? Explain your answer.

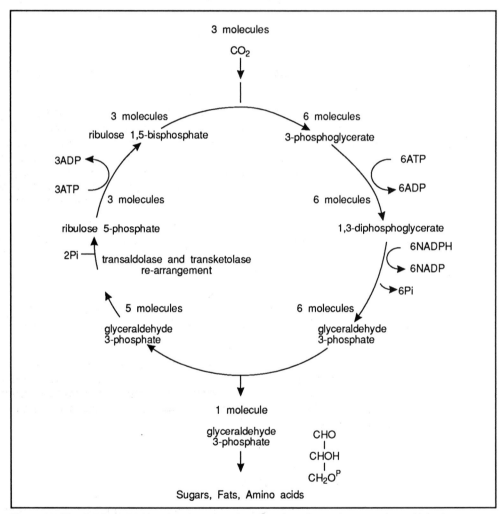

Figure 5.7 The Calvin cycle. Note that there are a large number of steps between glyceraldehyde 3-phosphate and ribulose 5-phosphate; these are omitted for clarity. (These have been described in greater detail in the BIOTOL text 'Energy Resource Utilisation in Cells').

Ⅱ Start at the top of this figure with 3 molecules of CO_2 and 3 molecules of ribulose 1,5-bisphosphate. What is the net effect of a full turn of this cycle? (Go round the cycle and see what has been used or produced at each stage)?

The net effect of the operation of the Calvin cycle is that three molecules of CO_2 are reduced to one molecule of glyceraldehyde 3-phosphate (3-phosphoglyceraldehyde) with the consumption of nine ATP and six NADPH molecules. Thus three ATP and two NADPH are used for each CO_2 molecule. There are two important points here. One is to remember that the ATP and NADPH used to reduce CO_2 can only come from the light reaction. The second is that ATP and NADPH are not consumed in equal amounts. Thus the plant must be able to generate them in unequal amounts.

5.3.11 What happens to the products of photosynthesis?

Plants carry out photosynthesis only in their shoots. Roots do not carry out this process and must be provided with food materials from the shoot. This is transported as sucrose.

Π Where is sucrose made in plant cells? (You have several choices: nucleus, mitochondria, chloroplast and cytosol).

export of fixed carbon

The answer is in the cytosol of leaf cells. Some of the 3-phosphoglyceraldehyde produced in the chloroplast is transported into the cytosol where a series of reactions take place leading to sucrose formation. The export of 3-phosphoglyceraldehyde is controlled by a specific carrier in the inner chloroplast membrane, the outer membrane being freely permeable to small molecules. Not all the 3-phosphoglyceraldehyde is exported. Much of it is used to produce the storage carbohydrate, starch, which is insoluble and forms starch grains. The 'fixed' carbon products of photosynthesis are sometimes referred to as photosynthates.

photosynthates

SAQ 5.7

Can you think what might happen to a chloroplast which, instead of making starch, simply stored excess 3-phosphoglyceric acid as the free acid?

SAQ 5.8

Photosynthesis can be said to be the conversion of light energy into chemical energy. What do you consider to be the first stable form of chemical energy in the process?

The sugar produced in photosynthesis is used as a food material by non-photosynthesising cells. Furthermore, starch and sucrose are common food materials used by animals. The energy of these compounds and that of fats can be released in reactions involving mitochondria so we will now examine this organelle and these processes.

5.4 The mitochondrion

5.4.1 The mitochondrion contains an outer and an inner membrane

The mitochondrion, of which there may be 100 to 2000 per cell, is bounded by two membranes with a space in between (Figure 5.8).

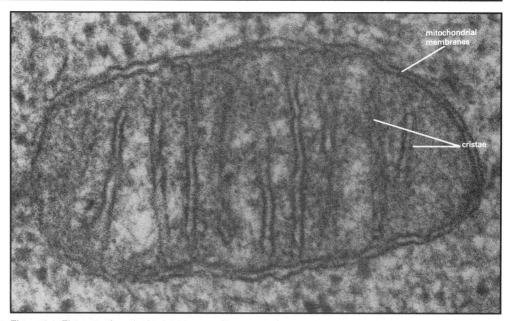

Figure 5.8 Electron micrograph of a mitochondrion (x 162 000).

cristae   Figure 5.8 shows that the inner membrane has extensions called cristae which stretch from one side of the mitochondrion to the other.

Π   Suggest two shapes which might produce such images and explain how you would decide which was correct.

The shapes seen could have been produced by cross sections of flat sheets or longitudinal sections of tubular extensions, both running across one side to the other. Tubular extensions are perhaps less likely because they would be expected to fall out of the plane of the section and, therefore, not give the continuous image seen in Figure 5.8. However, the question could be settled by examining serial sections.

In fact serial sections reveal a number of forms for the cristae, including those which are tubular and even triangular in cross section. The commonest form, however, is flat sheets as shown in Figure 5.9.

inner and outer   These studies show that four zones can be recognised in mitochondria. These are: the
membranes   outer membrane, the intermembrane space, the inner membrane and the central space
    referred to as the matrix. Studies show that the outer membrane contains large numbers
inter-membrane   of copies of a transport protein which form large aqueous channels through the lipid
space   bilayer. These allow molecules up to 10,000 daltons to pass through. The inner
    membrane, however, is largely impermeable. Thus while the solution in the
matrix   inter-membrane space is likely to be quite similar to that of the cytosol, the matrix space
    is quite different.

Mitochondria are the major components of a system which extracts the energy from organic nutrients so we will first of all see how this system fits into the broader scheme of organic nutrient breakdown.

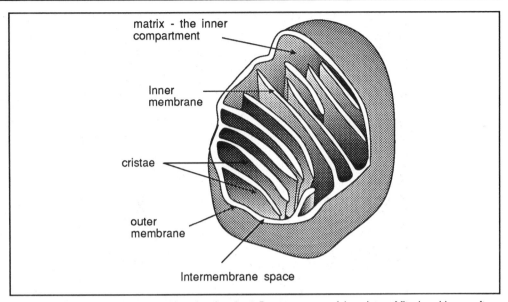

matrix - the inner compartment

Inner membrane

cristae

outer membrane

Intermembrane space

Figure 5.9 Diagram of a mitochondrion showing the 3-D arrangement of the cristae. Mitochondria are often shown as being sausage shaped, but time lapse micro-cinematography show that they bend and stretch, showing considerable variation in shape.

5.4.2 The breakdown of sugars can be aerobic or anaerobic

We will use glucose as our starting material; this can be easily obtained from sucrose or starch. The aerobic pathway, which requires oxygen, and the anaerobic pathway, which does not, are common over their early stages (Figure 5.10) producing pyruvic acid. This text is not primarily concerned with the details of metabolism, so we will not examine all of the reactions in detail. These are discussed in the BIOTOL text 'Principles of Cell Energetics'. We will however consider the broader principles involved.

The conversion of glucose to pyruvate shows a net production of two molecules of ATP and this is the sum total if sugar metabolism is being carried out in the absence of oxygen. Under these conditions, pyruvic acid may be converted to lactic acid, or ethanol and CO_2 or some other simple organic chemicals. In the presence of O_2 however, pyruvate is completely degraded to CO_2 and H_2O with the release of 36 more molecules of ATP (ie 38 ATPs in all).

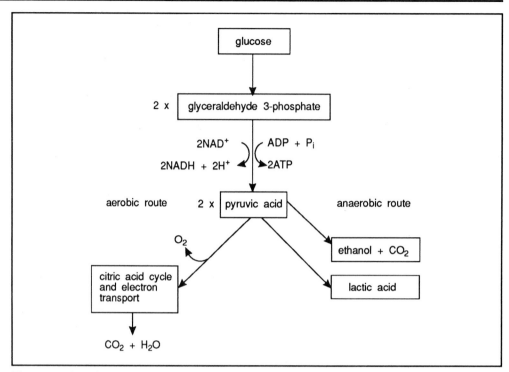

Figure 5.10 Overview of aerobic and anaerobic breakdown of glucose. We will see later how 36 ATP are generated in aerobic respiration.

∏ We have said that 2 molecules of ATP are produced when one molecule of glucose is converted to 2 molecules of pyruvate. Examine Figure 5.10 again. What else is produced?

You will be able to check your answer in the next section.

5.4.3 Metabolism of sugars in anaerobic conditions is inefficient

NAD⁺ is
reduced

The breakdown of glucose produces two molecules of pyruvate, a net production of 2ATP and the formation of two molecules of reduced nicotinamide adenine dinucleotide (NADH). As we will see later there is considerable energy available in NADH but it is not released under anaerobic conditions. Its fate is shown in Figure 5.11.

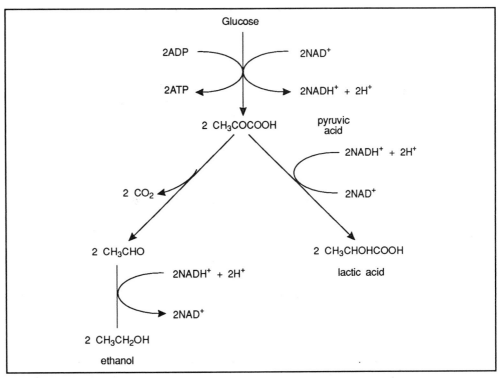

Figure 5.11 Examples of the reactions leading to reoxidation of NADH under anaerobic conditions.

SAQ 5.9

The amounts of cofactors present in the mitochondrion are low and this includes NAD^+. Once all the NAD^+ is reduced to NADH any reaction requiring NAD^+ will stop.

By examining Figure 5.11, explain how NAD^+ is made available for the breakdown of glucose during anaerobic metabolism.

end products still contain much energy

Ethanol production, as you know, is carried out by yeast. Lactic acid production occurs, for example, in muscle tissue at times of heavy loading and in the bacteria which make cheese. In these cases energy release is poor because the substrates are not fully broken down. There is still considerable energy in ethanol; indeed it is used as a fuel in some countries. The same is also true of lactic acid. Yeasts excrete ethanol but muscle tissue releases lactic acid to the blood stream and it is taken in by the liver and used in other reactions. Thus lactic acid is not wasted and its production allows for the continuous production of ATP under conditions of low oxygen and heavy loading.

Note that some micro-organisms produce other organic products when growing under anaerobic conditions (eg butyric acid, propionic acid). In each case, the basic metabolic reasoning is the same, namely to use the pyruvate produced during sugar catabolism to oxidise NADH (These processes are examined in detail in the BIOTOL text 'Principles of Cell Energetics').

SAQ 5.10

Can you suggest an occasion in your own body when your muscles might make lactic acid?

One of the factors which improves immensely with physical training is the muscle's ability to prevent the formation of lactic acid. If it builds up it causes muscle fatigue or cramp.

5.4.4 Pyruvate is fully oxidised by the operation of the citric acid cycle

In the presence of oxygen, pyruvic acid moves into the mitochondrion where, in the matrix solution, it is converted to a compound called acetyl coenzyme A (acetyl CoA) with the release of CO_2 and the formation of a molecule of NADH. Acetyl Co A can be looked upon as a form of activated acetate for the acetyl group can be transferred to a four carbon compound, oxaloacetic acid, forming a six carbon compound, citric acid. Citric acid gives its name to a metabolic cycle, also called the tricarboxylic acid cycle or the Kreb's cycle, in which the two carbon moiety of acetyl CoA is broken down in a step wise manner, thereby enabling the capture of the energy released. Figure 5.12 is a summary of the reactions of the cycle, and depicts only the major events. We will not be looking at the process in detail but it is worth mentioning here that acetyl CoA is also formed in the breakdown of fat. Essentially each turn of the cycle takes in one acetyl (CH_3CO-) group in the form of acetyl CoA and converts it to two molecules of carbon dioxide. Therefore, in terms of changes to the carbon compounds, we can represent each turn of the cycle as:

(margin notes: Kreb's cycle; tricarboxylic acid cycle)

$$CH_3CO- \; \rightarrow \; 2CO_2$$

(obviously the acetyl group is being oxidised and something else has to be reduced. In fact NAD^+ and FAD may be reduced).

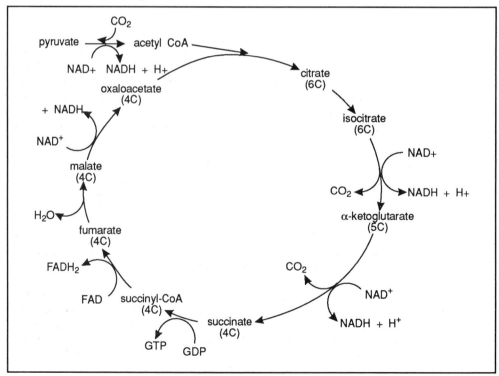

Figure 5.12 Simplified scheme of citric acid cylce (Kreb's cycle).

∏ Using Figure 5.12, see if you can calculate the product in terms of molecules of ATP, NADH and $FADH_2$ formed when the pyruvate derived from a molecule of glucose is completely degraded to three molecules of CO_2. Do the calculation yourself before reading on. Note that GTP can be converted to ATP by the reaction:

$$GTP + ATP \rightarrow ATP + GDP$$

The answer is that 2ATP, 8 NADH and $2FADH_2$ are produced. Do not forget two molecules of pyruvate are formed from each glucose and that an NADH is formed when pyruvate is converted to acetyl CoA.

∏ As another exercise, let us now include the reaction between glucose and pyruvate. What is the total now?

The total is 4ATP, 10NADH and $2FADH_2$. Do not forget that the NADH which is reoxidised in the process forming ethanol or lactate has not been oxidised if the pyruvate is passed into the Kreb's cycle and, therefore, contributes to the total.

For these calculations, we have made some simplifications as to what actually happens in cells. There are energy costs in shuttling these agents into and out of the mitochondria. A more advanced treatment of energy metabolism is dealt with in the BIOTOL text 'Principles of Cell Energetics'.

5.4.5 When does oxygen come into the picture?

The reactions of the Kreb's cycle described above are considered part of aerobic respiration but so far molecular oxygen has not been involved. This will now change. In the final series of reactions, carried out by components on the inner mitochondrial membrane, hydrogen atoms removed from NADH and $FADH_2$, are combined with molecular oxygen forming water.

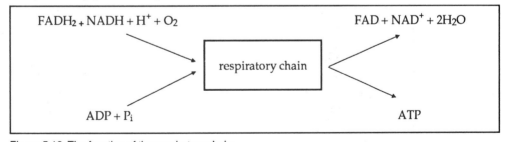

Figure 5.13 The function of the respiratory chain.

oxidative phosphorylation

respiratory chain

Since ATP is generated in this reaction involving molecular oxygen it is called oxidative phosphorylation. In this process the hydrogen atoms are separated into hydrogen ions and electrons. The electrons pass along an electron transport chain, called the respiratory chain, while the hydrogen ions are released into the aqueous matrix. They come together again only at the very end of the pathway where they recombine with each other and with oxygen to form water (Figure 5.13).

Thus NADH and $FADH_2$ are oxidised indirectly by molecular oxygen thereby refurbishing the supply of NAD^+ and FAD needed for the Kreb's cycle. In the absence of oxygen, the NADH and FADH$_2$ remain reduced and those reactions of the Kreb's cycle which rely on oxidised NAD$^+$ and FAD will stop. Thus the whole cycle will stop and anaerobic metabolism will take over.

5.4.6 Protons are pumped into the intermembrane space as electrons pass along the respiratory chain.

The components of the respiratory chain are arranged as in Figure 5.14 because it satisfies the energy requirements dictated by the redox potentials.

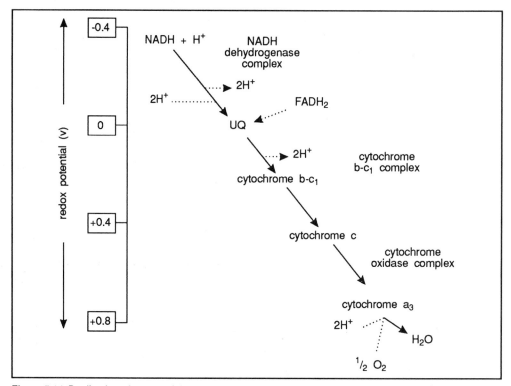

Figure 5.14 Decline in redox potential as electrons flow along the respiratory chain. UQ is ubiquinone, a hydrogen acceptor with similar properties to plastoquinone in chloroplasts. Only the main electron carriers are shown.

We met some similar components in the electron transport chain used in photosynthesis. Thus UQ (ubiquinone) is a quinone and can be oxidised and reduced in a manner described for plastoquinone in photosynthesis. Structurally these two are quite similar.

FMN (flavin mononucleotide) and FAD (flavin adenosine dinucleotide) are quite strong reductants and can supply the e$^-$ and H$^+$ ions for the reduction of ubiquinone.

It can be demonstrated also that as electrons are transported along the chain protons are moved from the matrix to the intermembrane space. Thus there is a need for organisation here just as there is on the thyllakoids of chloroplasts. Figure 5.15 shows an arrangement that would satisfy the requirements for proton pumping.

Let us follow the electrons down this chain. Begin with NADH. This reduces flavin mononucleotide (FMN) which shuttles across the membrane to reduce ferric ions in iron-sulphur centres. The ferric ions take up electrons from the reduced FMN and protons move into the intermembrane space of the mitochondria. The reduced iron-sulphur centres pass their electrons onto ubiquinone (UQ) which takes up two protons from the matrix of the mitochondria. The reduced ubiquinone (ubiquinol) gives up its electrons (to cytochrome b) and its protons to the intermembrane space. The reduced cytochrome b passes its electrons to other cytochromes. Ultimately cytochrome a_3 is reduced. This is oxidised by O_2. Protons are taken up at the same time.

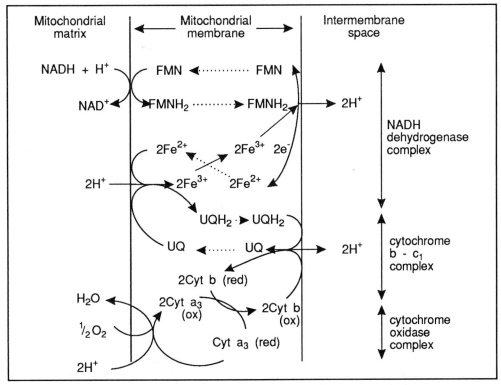

Figure 5.15 Postulated arrangement of components of the respiratory chain on the inner mitochondrial membrane. Fe^{3+} = ferric ions in iron-sulphur centres. UQ = ubiquinone. Note some cytochromes have been omitted for simplicity.

5.4.7 A membrane potential gradient co-operates with proton gradient to generate ATP

proton   Examine Figure 5.15 closely, you will see there is a net outflow of protons. This is used to drive the mitochondrial ATP synthetase situated in the inner membrane in a process analogous to that in chloroplasts.

SAQ 5.11

Dinitrophenol is a compound which has been useful in the study of the respiratory chain. It participates in the following reaction:

When exposed to high pH values (low H^+ concentration) dinitrophenol is charged and is lipophobic. At low pH the oxygen has a proton attached and the molecule is now lipophilic. What effect do you think this compound might have on ATP synthesis by mitochondria?

uncouplers

That was a difficult question. Do not worry if you did not solve it. Compounds such as dinitrophenol are called uncouplers of respiration because they uncouple the production of ATP from the operation of the respiratory chain.

symport

antiport

The proton gradient is not used only to generate ATP. Because it constitutes an energy source it can be used directly to drive active transport of, for example, pyruvic acid or phosphate into the matrix. In these two examples pyruvic acid or phosphate are transported against their concentration gradients by being co-transported with protons, which are moving down their gradient. This type of transport is referred to as a symport because both components are moving in the same direction. Antiport systems move compounds in opposite directions. We have represented these in Figure 5.16. The transport of ATP out of the mitochondrion occurs by an antiport system. The inner membrane of the mitochondrion is impermeable to ADP and ATP individually but a protein complex is present in the membrane which will exchange one for the other. For each ADP molecule moving in an ATP molecule moves out. This system is very important. If the energy of ATP was used up in moving it out of the mitochondrion into the cytosol there would be no mechanism for making ATP available outside the mitochondrion.

5.4.8 The ATP generating systems in mitochondria and chloroplasts are very similar

submito-chondrial particles

When mitochondria are broken open by suitable methods it is possible to isolate submitochondrial particles which are broken cristae that have resealed into small vesicles about 100nm in diameter. Electron micrographs of such particles show the presence of small spheres attached to the membrane by short stalks. We have presented a representative drawing of these in Figure 5.17. These particles can oxidise NADH by using O_2 and can generate ATP from ADP and inorganic phosphate.

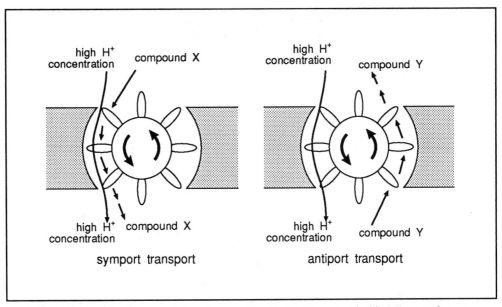

Figure 5.16 Highly stylised representation of symport and antiport transport. In this representation, we have chosen to show the concentration of protons as the driving force which turns a wheel rather like water running over a water wheel. The reader should, however, be alerted to the fact that this is merely a representation of the process - we are not suggesting that membranes are full of paddle wheels!

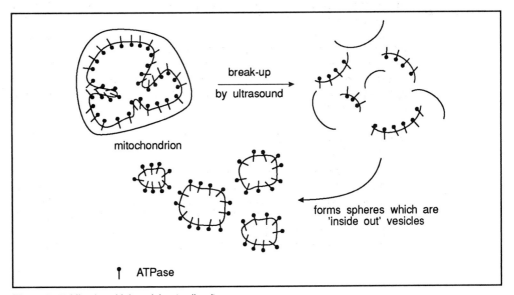

Figure 5.17 Mitochondrial particles (stylised).

The spheres can be easily stripped from the particles which results in the loss of their ability to form ATP. If the separated spheres are replaced and the particles reconstituted, ATP synthesis is regained. Subsequent work showed that the spheres and stalk are attached to a large transmembrane proton channel. ATP synthesis is therefore very similar to that found in chloroplasts.

5.4.9 Aerobic respiration is very efficient

During oxidative phosphorylation of NADH three molecules of ATP are formed (Figure 5.18). The electrons on $FADH_2$ are at a lower energy level and only two ATP are formed when it is oxidised.

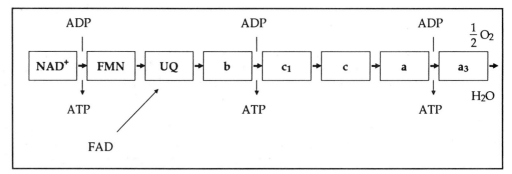

Figure 5.18 Oxidative phosphorylation associated with electron transport. Note b, $c_1 \rightarrow c \rightarrow a \rightarrow a_3$ are all cytochromes.

Table 5.1 shows a balance sheet for ATP production for the complete oxidation of a molecule of glucose.

		ATP
net ATP from glycolysis (substrate level phosphorylation)		2
NADH from glycolysis	2NADH	6 (4)
NADH from conversion of pyruvate to acetyl CoA	2NADH	6
NADH and 2 x $FADH_2$ from citric acid cycle	6NADH 2 x $FADH_2$	18 4
GTP from citric acid cycle $\equiv$ ATP (substrate level phosphorlyation)		2 ——
	Total	38

Table 5.1 Energy yield from the complete oxidation of glucose. Note two molecules of pyruvate are produced from each molecule of glucose.

Note that we have assumed that three molecules of ATP can be produced from each molecule of NADH which is generated within the mitochondrion. There is however, one proviso we must make. The ratio of 3 molecules of ATP per NADH oxidised seems only to be strictly true for NADH produced in the mitochondrion. NADH cannot pass from the cytosol into the mitochondrion. Therefore it has to pass its reducing potential into the mitochondrion. We will not deal with the details here, but in some cases the net effect of transporting the reducing potential of one molecule of NADH into the mitochondria followed by its oxidation, is to produce a net gain of 2 rather than 3 molecules of ATP.

Thus although we have calculated 6 ATP molecules being produced linked to the oxidation of the NADH generated by the conversion of glucose to two molecules pyruvate, this may not be strictly true. We can only guarantee 3 molecules of ATP for each NADH oxidised if the NADH is produced in the mitochondrion.

ATP produced directly in glycolysis and in the citric acid cycle are called substrate level phosphorylation and are not produced from proton gradients. Compare the yield with this system to that with oxidative phosphorylation which is linked to an electron transport system.

If glucose is burned in a calorimeter it yields 2870 kJ/mole of heat. The energy stored in the energy-rich ATP molecule is 30.5 kJ/mole. Thus 38 x 30.5 equals 1159 kJ are stored in 38 moles of ATP. This constitutes 1159/2870 or 40% of the available energy, which is much more efficient than any man-made machine.

SAQ 5.12	We have just calculated that 38 molecules of ATP can be formed when a single molecule of glucose is fully oxidised. If you actually determine the number of ATP molecules formed by cells when they respire glucose aerobically the figure is usually much less. What reasons can you suggest for this?

SAQ 5.13	In anaerobic respiration in yeast, acetaldehyde accepts hydrogens from NADH, so allowing glycolysis to continue. Acetaldehyde is said to be the final acceptor of hydrogen. What is the final acceptor of hydrogen in aerobic respiration? Are there any advantages in using the final acceptor you have suggested over using acetaldehyde?

5.5 Mitochondria and chloroplasts have probably evolved from endosymbiotic bacteria

endosymbiont

Mitochondria and chloroplasts contain DNA arranged as a circular molecule and this is the situation in prokaryotes. The endosymbiont hypothesis holds that eukaryotic cells originally contained no chloroplasts or mitochondria but then established a stable relationship with a bacterium, which conferred on them the ability to carry out oxidative phosphorylation. The bacterium evolved into the mitochondrion. Animal and plant mitochondria are so similar in structure and function that this primary event of symbiosis is considered to have occurred before plants and animals embarked on their separate evolution. Later the establishment of symbiotic relationship with an organism resembling present-day cyanobacteria, which are photosynthetic, gave rise to chloroplasts. Although mitochondria and chloroplasts contain DNA they do not now make all of the proteins that they need. Indeed, most of the genes coding for mitochondrial and chloroplastic proteins are contained within the nucleus suggesting extensive gene transfers, thus giving the nucleus overall control. However, some proteins *are* made in the two organelles, on ribosomes similar in organisation to prokaryote ribosomes. There are also many similarities in the proteins of the organelles and prokaryotes and in the lipids and pigments they contain. Thus this proposal, which when first made seemed like a theory from a science fiction fantasy, is now generally accepted.

Having raised the question of the expression of genetic information in mitochondria and chloroplasts perhaps we should now examine this in cells as a whole. This will be the topic of the next two chapters.

SAQ 5.14

Which of the following applies to zones 1, 2, 3, 4, 5 and 6 in the diagrams below?

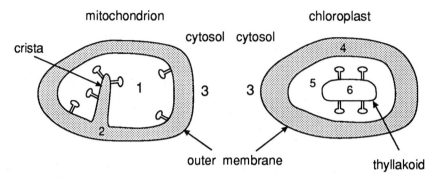

a) It has a pH of 5.

b) It has a pH of 7.

c) It has a pH of 8.

d) ATP is synthesized here.

e) Glycolysis occurs here.

f) Sucrose synthesis occurs here.

g) Starch synthesis occurs here.

h) Pyruvate is converted to acetyl CoA here.

Summary and objectives

In this chapter, we have described the structure and occurrence of chloroplasts and described the processes by which light energy is captured and transformed into chemical energy (ATP) and reducing potential. We have also examined how these two products are used to convert carbon dioxide into organic compounds.

We have also examined the other major biological energy generating system - the mitochondrion. We have placed the function of the mitochondrion within the overall context of the breakdown and oxidation of organic substrates using glucose metabolism as our specific example. We have described the structure and function of mitochondria and calculated the efficiency of energy harvesting from chemical oxidation using mitochondrial processes. This chapter should have revealed a similar mechanism for ATP synthesis coupled to electron transport in both chloroplasts and mitochondria.

Now that you have completed this chapter you should be able to:

- define the terms autotroph, chemotroph, phototroph and heterotroph and point out similarities and differences between them;

- demonstrate an understanding of how the fractionation of organelles aids the elucidation of the functions of their parts;

- describe similarities and differences between the structure of chloroplasts and mitochondria and comment on their possible evolutionary origin;

- define the terms absorption spectrum and action spectrum and show how they aid the study of photosynthesis;

- show an understanding of and describe in general biochemical terms the light dependent reactions and light-independent reactions of photosynthesis;

- define and describe in general biochemical terms aerobic respiration;

- arrange redox reagents in a logical sequence in accordance with their redox potentials;

- explain the difference between substrate level phosphorylation, cyclic photophosphorylation, non-cyclic photophosphorylation and oxidative phosphorylation.

The expression of genetic information: I - the nucleus

The expression of genetic information: I - the nucleus

6.1 Introduction

central role
of DNA in
specifying cell
activities

It is a property of multicellular organisms that they possess cells which differ from each other in structure and function. The cell, of course, generates its own structure. It does this by first making enzymes, some of which catalyse the reactions which produce building blocks, others of which catalyse the putting together of these building blocks to form the specific structure of the cell. The production of a cell with a particular shape and size is only part of the process, however, for cells of multicellular organisms also differ in the types of biochemical reactions which predominate within them and these too are catalysed by specific enzymes. Thus we can see that the basic difference between cells of different structure and function is their complement of enzymes. Some of these are responsible for generating the specific nature of the cell's structure while others specify the types of reactions which take place on a day to day basis. The compound which carries each cell's specific blueprint is DNA. Thus the study of the expression of genetic information becomes one of studying the production of enzymes and structural proteins following the instructions layed down in a cell's DNA. The DNA resides mainly in the nucleus, whereas the process of enzyme (protein) synthesis occurs in the cytoplasm. So we will begin our story by looking at the nucleus and then follow the trail out into the cytoplasm. We should remember that a small amount of DNA is also present in the two major organelles, the mitochondrion and the chloroplast.

6.2 The nucleus

6.2.1 The nucleus has a double membrane with holes in it

nuclear
envelope,
perinuclear
space, nuclear
pores

The nuclei from eukaryotic cells are thought to be surrounded by a double membrane referred to as the nuclear envelope. The space between the membranes is referred to as the perinuclear space. Evidence from freeze-fracture electron micrographs indicates that there are numerous pores (nuclear pores) within these membranes. The external and internal membranes of the nuclear envelope come together at the nuclear pores, thus sealing the perinuclear space (the space between membranes).

6.2.2 Nuclear pores are more than just holes

pore complex

You will probably agree that the possession of a double membrane could pose severe transport problems for the nucleus. This problem, however, is considered to be solved by the nuclear pores which are thought to function in the exchange of materials between the nucleus and the cytoplasm. The pore is not, however, simply a 90-110nm hole. A number of large protein granules are present on the edge of the pore forming an octagon. These proteins partially block the pore leaving a narrow water-filled space running from the cytoplasm to the nucleus.

SAQ 6.1

A series of fluorescently labelled proteins were prepared, separately injected into the cytosol and their distribution determined after various time intervals. Some were found to have migrated into the nucleus after the following times:

protein size	time to enter nucleus
5000 daltons	30 secs
17 000 daltons	2 mins
44 000 daltons	30 mins
66 000 daltons	1 hr

Would you expect results of this sort from active transport or endocytosis? If not what interpretation can you put on this data?

Experiments such as those described in SAQ 6.1 suggest that the pore is water-filled and has an open channel with a diameter of about 9nm.

6.2.3 What is the role of the nuclear pore complex?

selective movement of molecules into and out of the nucleus

As we will see in later sections there is a considerable movement of molecules through the nuclear pores which, in a typical mammalian cell, number between 3000 and 4000 per nucleus. Most of the molecules which migrate are small and diffuse through easily. Many others are not, however. The enzymes DNA and RNA polymerase which are involved in nucleic acid synthesis, are between 100,000 and 200,000 daltons in molecular mass and they are manufactured in the cytoplasm. On the other hand the ribosome subunits, which are used in the cytoplasm for protein synthesis, are made in the nucleus. Ribosomal subunit particles are about 15nm in diameter. Electron micrographs of nuclear pores often show them blocked by electron dense particles which stain positively for ribonucleoprotein, the major component of ribosomal subunits. These observations are interpreted as suggesting that the large molecules blocking the pores are ribosome subunits in transit and that in some way the granular proteins of the pore complex help this process. Note that the nuclear pore without its surrounding protein complex is sufficiently wide to allow the 15nm ribosome subunit to pass through.

∏ Write down an additional property in relation to transport that the pore complex might have.

We have noted two type of molecules which need 'help' to get through the pore. The key question is, whether or not they are selectively transported. The pore complex could conceivably have the property of recognising molecules and deciding which to let through.

∏ Why bother with a pore complex? Why not simply have an open pore? Write down a few ideas of what you think might happen.

We do not know what would happen if the pore was open but we can suggest some possibilities. Much larger molecules could diffuse in or out easily. This might let in compounds which would disrupt operations inside the nucleus. Conversely, compounds required in the nucleus might escape.

6.2.4 The nuclear envelope is connected to the endoplasmic reticulum

We have seen that the outer and inner membranes come together at the nuclear pores. In addition the outer membrane is continuous with the endoplasmic reticulum which means that the perinuclear space is continuous with the lumen within the endoplasmic reticulum. This is shown diagrammatically in Figure 6.1.

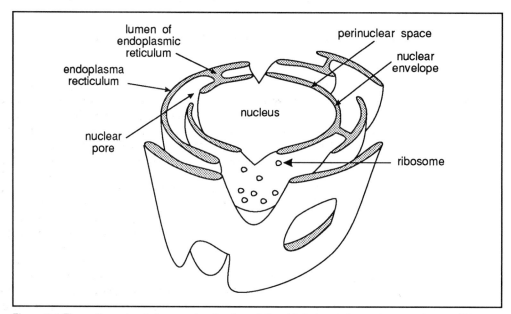

Figure 6.1 Three-dimensional diagram showing the relationship between the nuclear envelope and the endoplasmic reticulum. Note the ribosomes on the outer nuclear membrane.

nuclear envelope and endoplasmic reticulum exchanges

It can be seen that the perinuclear space is part of a much larger compartment that spreads throughout the cell. This is not the only consequence. It is considered that membrane can be exchanged between the nuclear envelope and the endoplasmic reticulum. Thus nuclear membrane area can increase by the movement of membrane from the endoplasmic reticulum to the outer membrane and from there to the inner membrane through the nuclear pore. The reverse can also occur. What happens to the pore complex while this is going on? We do not know the answer to this question but if the pore complex has an important controlling role it would be suprising if it was continually being dismantled and reassembled. Membrane exchange occurs mainly during preparation for and recovery after mitosis (nuclear division as part of cell division). Since the whole nuclear membrane disappears during mitosis the role of the pore complex in controlling movement of molecules into and out of the nucleus ceases to have meaning at this time.

6.2.5 The nucleus has its own skeleton

nucleoplasm

Many electron micrographs of the nucleus show an electron-dense layer just inside the inner membrane on the nucleoplasm (nuclear matrix) side. This material has been isolated and shown to form a fibrous network made up of mainly three proteins. This network forms a layer towards the periphery of the nucleus and is considered to be attached to the inner membrane by binding to specific integral membrane proteins. It is

nuclear lamina

called the nuclear lamina and is considered responsible for maintaining the shape of the nucleus. Furthermore it is connected to the granular proteins in the nuclear pore. Thus the nucleus has its own skeleton.

| SAQ 6.2 | The nuclear lamina varies in thickness in different sections and sometimes appears to be absent. Can you suggest a very sensitive method for detecting it? |

chromatin fibres   Finally there is evidence that the chromatin fibres, (which contain the cell's DNA), are attached to the nuclear lamina. We will now examine the organisation of chromatin.

6.3 Organisation of the genetic material

6.3.1 Chromatin consists of DNA and protein

chromosomes   The DNA of prokaryotes consists of a single circular double-stranded molecule whereas that of eukaryotes is organised into a number of separate chromosomes each of which contains a single linear double-stranded DNA molecule. We will not discuss the organisations of DNA in prokaryotes any further here, but focus on the DNA in eukaryotes. We remind you that in prokaryotes DNA is in the form of a circle which is twisted up and packed into cells probably using polyamines, proteins and RNA to aid packing. The DNA of eukaryote organisms is associated with a variety of proteins some of which have a structural role while others function as enzymes. The structural proteins

histones   belong to a class called histones which are relatively small with a high proportion of basic amino acids such as arginine and histidine. At normal cellular pH values, these amino acids have positively charged side chains which enable them to bind tightly to

chromatin   DNA. The term *chromatin* is given to this association of DNA and histone and the association appears to be of a permanent nature. The histones are composed of five types of protein but here we do not need to know much more of the details of their structures. They are however basic proteins and carry a positive charge at normal cell

conservation of structure   pH values. Thus they are electrostatically attracted to the nucleic acids. One further point is worth making. Four of the histones show remarkable conservation of structure. This means that histones from different organisms, such as cows and peas, show very little difference in their amino acid sequence, suggesting that they have crucial functions depending upon this precise sequence. The other proteins associated with DNA are non-structural and are grouped as non-histone proteins.

It is estimated that each DNA molecule in a eukaryote would be around 5cm long if it was extended. During the process of mitosis, however each molecule condenses to about 5µm in length. This requires considerable folding and organisation.

6.3.2 DNA and histones form nucleosomes, the fundamental unit of a chromosome

nucleosome   The nucleosome consists of a core of histone molecules with DNA wound around each core, the DNA molecule serving to hold adjacent nucleosomes together. This arrangement is referred to as the 'beads on a string' arrangement (Figure 6.2).

The nucleosomes can be separated from each other by treatment with enzymes which degrade DNA molecules. These are called deoxyribonucleases or DNases. We can also note here that enzymes which degrade RNA are ribonucleases (RNases) whereas enzymes which can degrade either are called nucleases. Treatment with DNase for a short period digests the linker DNA whereas that wound around the histone is protected. The beads can be separated from each other and when the DNA is dissociated from the histones it is seen to consist of 146 basepairs. The histones form a disc-shaped core about 11 nm wide, around which it is thought that the DNA can be wound twice. The linker DNA has about 60 basepairs so each bead and its linker contains about 200 basepairs.

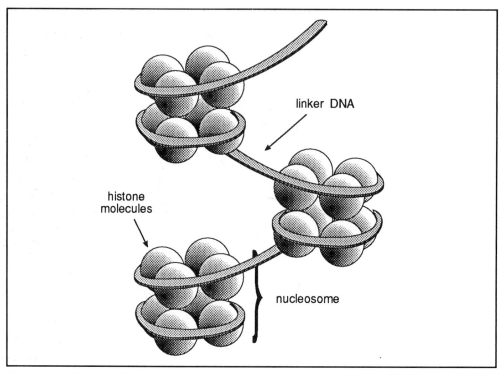

Figure 6.2 A diagrammatic representation of a partially decondensed piece of chromatin exhibiting the beads on a string organisation.

∏ Mark on Figure 6.2, the dimensions described in the previous paragraph.

SAQ 6.3   If the histone core of a nucleosome is 11nm in diameter and each linker about 20nm long what effect does the nucleosome have on the linear length of DNA?

The formation of nucleosomes begins the process of length reduction but not by anything like as much as the next level of organisation.

6.3.3 Beads on a string are also coiled

30nm chromatin fibre

When nuclei are very carefully broken open and observed under the electron microscope the chromatin appears in a form with a diameter of 30nm, known as the 30nm chromatin fibre.

histone H₁

Further treatment of the 30nm fibre gives rise to the beads on a string form showing that the 30nm fibre is some sort of coiled beads on a string form. Two proposed models for this coiling is shown in Figure 6.3 but the exact form is uncertain. Indeed the evidence suggests that more than one type may occur. What is known is that this coiling much more dramatically reduces the linear length of the DNA; by almost 20 fold, our 5cm DNA molecule now being down to about 1mm in length. A different histone molecule, called histone H_1 is involved in linking adjacent beads together in the 30nm fibre.

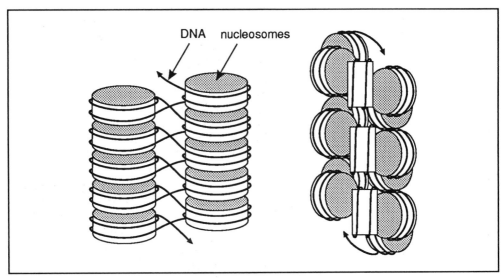

Figure 6.3 Models to show two suggestions for the packing of nucleosomes (highly stylised).

6.3.4 The squeeze goes on in looped domains and further coiling

looped domains

Remembering that a nucleus is approximately 5μm in diameter, the 1mm long by 30nm wide fibre still has a long way to go to fit in. The next folding system is referred to as the looped domain (Figure 6.4).

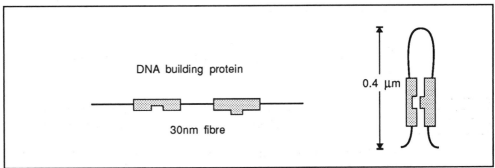

Figure 6.4 Folding of a 30nm fibre to make a looped domain.

polytene

lampbrush chromosomes

meiosis

interphase

Looped domains were seen first in unusual chromosomes found in certain insect cells, (eg the polytene chromosomes), and in egg cells, (eg the lampbrush chromosomes). Polytene is the name given to chromosomes formed by the production of many hundreds of copies of DNA, which remain bound together forming giant chromosomes which are visible in the light microscope. The lampbrush chromosome form is shown by certain egg cells undergoing preparation for reproductive cell division, meiosis. This preparative period may last for many years, during which time the chromosome is actively involved in directing RNA synthesis. Both polytene and lampbrush forms produce very striking looped domains which was why they were the first seen looped domains. It now appears that looped domains are present in 'normal' chromosomes and are responsible for a further 10 fold reduction in length. Our 5cm DNA molecule is now down to a length of 100μm. This is likely to be the form of chromosome present in an interphase nucleus (ie one which is not showing cell division).

condensed

During the process of mitosis and meiosis the looped domain form becomes further coiled, into a superhelix. This reduces the chromosome to its shortest and thickest, being less then 5μm long, and in this state it is referred to as being condensed.

SAQ 6.4

Which of the following statements apply to 1) nucleosomes, 2) 30nm fibres, 3) looped domain chromosomes, 4) condensed chromosomes?

a) Only produced during mitosis.

b) Contain a DNA double helix.

c) Describes the structure composed of DNA looped around histone cores.

d) Reduces the overall length of the DNA by about 10 fold.

e) Formation reduces the length of the DNA molecules by 20 fold.

f) Involve units held together by histone H1.

g) Structure was first seen in polytene and lampbrush chromosomes.

h) Can be seen in the light microscope.

6.3.5 The significance of DNA packaging

naked DNA

It should be clear from the above description that the DNA of eukaryotes is very carefully packaged. Although there is some evidence for the packaging of DNA in prokaryotic cells, it is in no way as complex as that described in eukaryotes. DNA in prokaryotes may be associated with protein, but such association is rather poorly defined and many authorities tend to regard prokaryotic DNA as naked.

replication

These differences have important consequences. To explain what these consequences are we would like to draw an analogy. Imagine that DNA was like a strand of wool. In prokaryotes the DNA is rather like an unwound ball of wool which has been gathered up into a rather random loose cluster. In eukaryotes the 'wool' is neatly wound into balls (chromosomes) which are kept together in a box (nuclear envelope). What are the consequences of the differences in packaging? Which form of wool (unwound or balls in a box) is easiest to handle? Of course the balls in the box. At various stages of the cell cycle but particularly at mitosis it is essential for the cell to 'manage' its DNA. First it has to be replicated (copied) and then the two copies have to be separated and each passed to a daughter cell. We will discuss this process in detail in a later chapter. For now we wish to make the point that it such easier to handle large quantities of DNA if it is suitably packaged. We all know how easy it is to get knots into unravelled wool!

Thus one of the consequences of the packaging of DNA in eukaryotes is that it enables eukaryotes to handle greater quantities of genetic information. At the end of Chapter 1 we made the point that multicellularity demanded considerable more genetic information than the unicellular state. Thus, at least in part the evolution of a DNA packaging system may have enabled cells to carry sufficient genetic information to enable multicellularity to develop.

The packaging of DNA and its separation into a distinct compartment (the nucleus) has other important consequences. In prokaryotes, where the naked DNA is suspended in the cytoplasm, it is readily accessible to those enzymes involved in gene expression. The wrapped DNA of eukaryotes is not so readily accessible. We will turn our attention to this problem by examining the steps involved in gene expression in eukaryotes.

6.4 Gene expression

In discussing gene expression here, we have assumed that you have a basic knowledge of the structure of DNA and RNA. If this is not the case, we would recommend the BIOTOL text 'The Fabric of Cells' which provides the necessary background.

6.4.1 Gene expression involves many steps

As we noted in the introduction to this chapter gene expression involves the manufacture of a set of proteins, many of which are enzymes. The blue-print for this is the DNA in the chromosomes but the proteins are actually made on ribosomes in the cytoplasm. Since the DNA stays in the nucleus there must be some way of getting the information to the cytoplasm. This is achieved by producing a copy of the DNA which is then transported to the cytoplasm. This copy is an RNA molecule and because of its function it is called messenger RNA (mRNA). The message 'written' in the DNA is copied in the mRNA. This process is called *transcription*. In the second part of the process the information in mRNA is converted into the amino acid sequence of a protein. This is obviously a change of form of the information and is called *translation* (Figure 6.5).

mRNA

transcription

translation

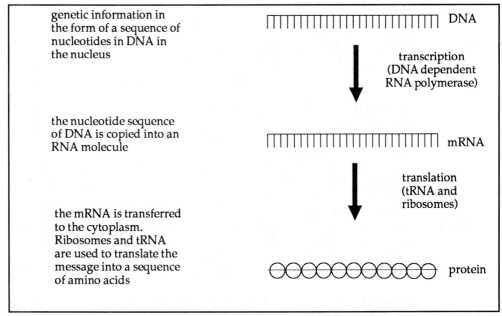

Figure 6.5 Overview of the transfer of genetic information from nucleus to cytoplasm.

The information in the mRNA molecule is contained within the sequence of its bases. Each amino acid in a protein is coded by a group of three bases on the mRNA, called a codon. In the process of protein synthesis the codon is recognised by another form of RNA called transfer RNA (tRNA) by its possession of a group of bases complementary to the codon, called the anticodon. The amino acid specified by the anticodon is attached to another part of the tRNA molecule and there are specific tRNA molecules for each amino acid. Because the tRNA enables the base sequence to be converted to an amino acid sequence it is sometimes referred to as a adapter molecule. The actual process of reading the message and lining up the tRNA molecules so that the amino acids can be joined together, takes place on the ribosomes, which can, therefore, be looked upon as protein synthesising machinery. Each ribosome consists of two different subunits, a large and a small subunit which are made of protein and a third kind of RNA, ribosomal RNA (rRNA).

tRNA

rRNA

codon

anticodon

adapter molecule

SAQ 6.5

Put a cross against each of the following statements if true for the nucleic acid listed.

	DNA	tRNA	rRNA	mRNA
1) Is involved in translation.				
2) Is an adapter molecule.				
3) Contains codons.				
4) Contains an anticodon.				
5) Is in the form of a double helix.				
6) Is involved in transcription.				
7) Contains the whole of the genetic code.				
8) Is made in the nucleus.				

6.4.2 RNA synthesis is catalysed by RNA polymerase

DNA dependent RNA polymerase transcriptase

In prokaryotes the synthesis of all RNA molecules is catalysed by a single RNA polymerase. This is also called 'DNA dependent RNA polymerase' or transcriptase. This recognises specific regions of DNA, the promoters or start regions, where synthesis of RNA starts, nucleotides being added one at a time from the 5' to the 3' end of the molecule. This process continues until a second specific region is reached, the termination signal, where the enzyme disengages from the DNA releasing the new RNA molecule. Do not forget that the RNA is made as a complementary copy of one of the DNA strands. Thus where the DNA contains guanine, the RNA polymerase will put cytosine into the RNA strand, and where the DNA contains adenine, the polymerase will add uracil to the growing RNA strand (see Figure 6.6).

Note that although RNA is synthesised from 5' to 3' ends the RNA polymerase reads the DNA strand from 3' to 5'. RNA synthesis is said to take place in an antiparallel manner.

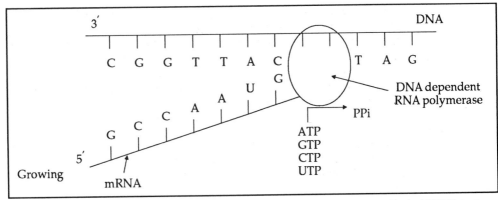

Figure 6.6 The structural relationship of DNA and RNA. Further details are provided in the BIOTOL text 'The Fabric of Cells'.

<table>
</table>

3 different RNA polymerases in eukaryotes

Also note that the incoming ribonucleotides are triphosphate nucleotides and that two high energy bonds are broken in attaching each ribonucleotide to the growing chain. The principle of RNA synthesis is the same in eukaryotes but there are three different polymerases. RNA polymerase II transcribes the genes that code for proteins; polymerase I and III code for RNA molecules used in ribosomes. Whereas purified bacterial RNA polymerase can be used to generate RNA molecules *in vitro*; this has only recently been achieved with eukaryotes and the precise requirements are not yet known. Needless to say the three eukaryotic polymerases have different start and termination signals and therefore they 'read' different parts of the DNA.

6.4.3 Transcription occurs on DNA in nucleosomes

transcription occurs on nucleosomes

Ordinary thin section electron micrographs show little useful detail about how genes are transcribed. Much more useful pictures are obtained if the nucleus is broken open osmotically so that its contents spill out onto an electron microscope grid. Close to the nucleus the chromatin is very dense and the detail cannot be resolved. Towards the edges of the chromatin, however, it is much more dilute and pictures can be obtained that show that transcription occurs on nucleosomes.

6.4.4 Transcription involves partly unwinding the helix

Usually only one of the two DNA strands of a protein coding region is transcribed. The other strand has a complementary base sequence which would code for a different sequence of amino acids. The strand that is transcribed is called the coding or sense strand of the DNA, the non-transcribed strand being the non-coding or nonsense strand. The non-transcribed strand is also called the antisense strand. It is somewhat mind-boggling to realise that a DNA strand may serve as a sense strand at one point and a nonsense strand at another. This decision, as it were, is governed by the location of the starter region; it can be present on either strand but will of course direct the synthesis only in the 5' to 3' direction. This point is illustrated in Figure 6.7.

sense and nonsense

DNA strands

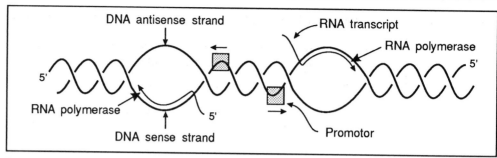

Figure 6.7 Transcription of RNA showing that one strand of DNA can act as a sense strand at one point and a nonsense strand at another (re-drawn from Villee, Solomon, Martin, Martin, Berg and Davis Biology, Saunders College Publishing).

In order for the DNA to exhibit its template function it is considered that the helix must unwind ahead at the point of insertion of the nucleotide. As the RNA molecule increases in length it actually begins to form a double helix with the DNA template, but this is less stable than the DNA-DNA helix which soon re-establishes itself. The unwinding does not involve breaking the DNA strand and, the evidence suggests that the core histone of the nucleosome remains in place during transcription. The organisation must allow the partial unwinding of the DNA helix and separation of the two strands so as to provide enough space for RNA polymerase to approach and bind to the DNA template. The magnitude of the puzzle here can perhaps be better appreciated when it is realised that RNA polymerases are about the same size as the histone core of a nucleosome.

ribonucleo-
proteins
As RNA molecules come off the production line they associate with proteins thus forming ribonucleoprotein particles. This is similar in principle to the packaging of DNA into nucleosomes, although different in detail.

6.4.5 mRNA molecules are cut down to size before being released from the nucleus

capped RNA
The RNA molecules produced by RNA polymerase II, those coding for proteins, are known as heterogeneous nuclear RNA (hnRNA). Before they are released to the cytoplasm three things happen to them. First the 5' end is capped by a structure that allows it to bind to a ribosome, and this occurs while the RNA is still being transcribed. Once transcription is complete and the RNA molecule is released another form of

poly A tails
polymerase add 100-200 residues of adenylic acid to the 3' end (poly A tails). The RNA molecules are now called primary RNA transcripts. HnRNA molecules contain 10 000-50 000 nucleotides but before they reach the cytoplasm this is reduced to 500-3 000.

introns
They still have a capped 5' end and a poly-adenosine chain at the 3' end. Extensive sections of nucleotides are removed from the middle regions of hnRNA while the two

exons
ends are simultaneously rejoined. The sections that are removed are called *introns*, the remaining parts are *exons* and the process of rejoining is called RNA splicing. The

splicing
situation just described, referred to as RNA processing, occurs in eukaryotes but not prokaryotes. Prokaryotes do not produce RNAs which possess introns, capped 5' ends or poly-adenylated 3' ends.

Certain evidence suggests that the removal of introns in eukaryotes is in some way linked to the process of mRNA release to the cytoplasm. Artificial genes can be created and their expression tested by insertion into cells. These genes are transcribed in the nucleus but, unless the gene originally possessed an intron, its mRNA product stays in the nucleus. In these cases the act of RNA splicing appears to generate an export signal. Other signals must exist, however, because not all eukaryotic genes have introns.

SAQ 6.6

Fill in the labels 1-5 of the diagram, using a selection of the words provided:

Stop signal, translation, intron, poly-adenylic acid zone (poly A tail), messenger RNA, 5′ cap, transfer RNA, start signal, exon.

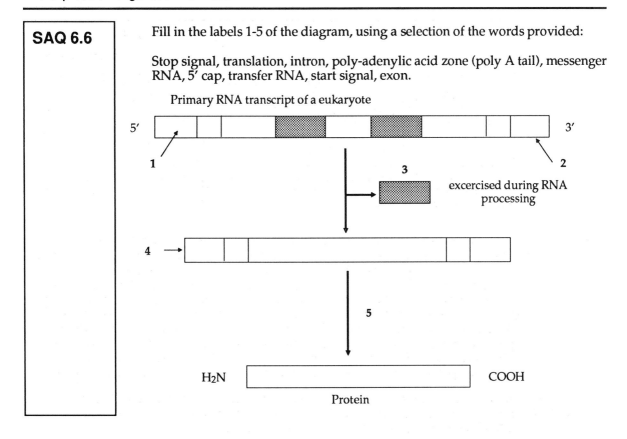

Primary RNA transcript of a eukaryote

6.5 Ribosome production

6.5.1 Each cell contains many copies of the genes coding for ribosomal RNA

multiple gene copies

A growing cell contains about 10 million ribosomes and each contains a large and a small subunit each consisting of protein and rRNA. With the synthesis of a protein from mRNA there is a considerable amplification factor if the mRNA molecule can be used several times. The most striking example of this is the blood cell protein haemoglobin. Here each mRNA produces as many as 10 protein molecules per minute, the total being of the order of 10,000 protein molecules per mRNA during the life of the cell. However, rRNA is the final product of the ribosome gene so if a cell needs lots of rRNA molecules in order to make a lot of protein it must either develop the ability to make them very quickly or produce multiple copies of the gene. The latter is the chosen path and human cells, for example, contain about 200 copies, distributed over five different chromosomes. The rRNA gene copies are arranged in tandem fashion on a chromosome and are separated by non-transcribed DNA called spacer DNA (Figure 6.8).

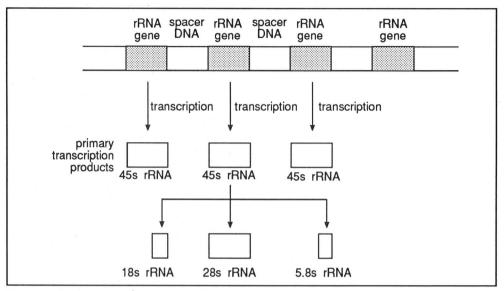

Figure 6.8 The tandem array of ribosomal RNA genes and the processing of the major RNA gene products (not to scale).

6.5.2 Ribosomal precursors are made in the nucleolus

Ribosomes consist of two subunits, each of which contain proteins and RNA. The ribosomal RNA is made from the ribosomal RNA genes. These genes are of two types with regard to the size of their product. We will deal with the large one first. The major rRNA genes are transcribed by RNA polymerase I and produce the primary RNA transcript known as 45s RNA (s is the Svedberg unit of sedimentation coefficient. It is a measure of the rate the molecule will sediment under centrifugal force). This molecule is degraded in the nucleus to three separate rRNA molecules the 18s, 28s, and 5.8s RNA molecules, the first of which contributes to the small ribosome subunit, the other two to the large subunit. The production of these three precursors from the same rRNA transcript ensures, of course, that they are produced in equal amounts. Another rRNA molecule, known as 5s RNA, is added to the 28s and 5.8s RNA which eventually form the large subunit. The 5s RNA is made from what may be called the minor rRNA genes. RNA polymerase III is used to transcribe these genes which are entirely separate from the 45s rRNA genes, although they are also present in multiple copies in tandem.

All of the various forms of RNA are packaged with protein as soon as they are synthesised. They then undergo a variety of processing events which ultimately produce the small and the large subunits of the ribosome. Only now do they leave the nucleus to join together to form the mature ribosome. The processing of the ribosome precursors occurs in the nucleolus, a large diffuse structure in the nucleus. The nucleolus is attached to chromosomes carrying rRNA genes. The specific DNA loop carrying the rRNA genes is referred to as a nucleolar organiser region. The production of the ribosome precursors is described diagrammatically in Figure 6.9. Begin at the top and work down to the final product. Note the nature and orgins of the material passed into the nucleus. You should realise that our drawing is somewhat stylised. The nucleolus does not normally more or less fill the nucleus as we have drawn it. We had, however, quite a lot to include in this diagram because ribosomal RNA processing is quite complex.

nucleolus

nucleolar
organiser

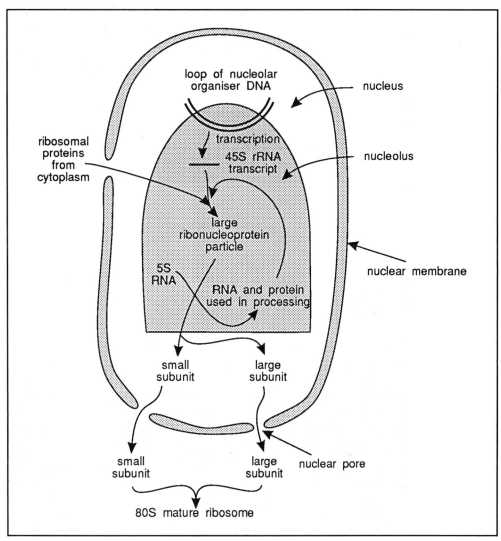

Figure 6.9 Diagram showing the involvement of the nucleolus in ribosome synthesis.

The finished ribosome is about half protein and half RNA by weight. The small subunit contains a single RNA molecule plus 33 different ribosomal proteins, whereas the large subunit consists of three RNA molecules and more than 40 different proteins.

The nucleolus is a somewhat unusual structure in that it changes its size according to the activity of the cell. Thus in dormant plant cells it is very small but can occupy up to 25% of the nuclear volume in cells actively synthesising RNA and protein. Furthermore its appearance changes during the cell cycle. As the cell approaches mitosis the chromosomes condense and the nucleolus gets smaller and eventually disappears. It reforms after division, initially being seen as several small nucleoli which quickly coalesce into one. Presumably a nucleolus reforms on each chromosome carrying rRNA genes.

SAQ 6.7

List the compounds that you would expect to find moving out of the nucleus to the cytoplasm.

SAQ 6.8

Name as many different proteins as you can which you would expect to find moving into the nucleus from the cytoplasm. What is their function?

We have now reached a suitable position to end this chapter. Having described the events taking place in the nucleus we are now ready to move out into the cytoplasm and continue the story of the expression of genetic information.

Summary and objectives

In this chapter we have learnt of the basic processes involved in the expression of genetic information. We explored the way in which the genetic information is packaged into chromosomes and the way these chromosomes are retained within the nucleus. We have also explored the nature of the nuclear membrane, its relationship with intra-cellular endoplasmic reticulum membrane and the regulation of the passage of materials between cytosol and the nucleus through the nuclear pores.

We have also examined the processes of converting nucleotide sequences of DNA into RNA and explored the processing of this RNA within the nucleus to produce either mRNA or ribosomes. We have discussed the role of the nucleolus in this process.

Now that you have completed this chapter, you should be able to:

- describe in general biochemical terms what is meant by the phrase "expression of genetic information";

- explain how chromatin differs from DNA;

- define the terms "beads on a string form of chromatin" and "30nm fibre" and show their relationship to the DNA double helix and the form of the chromosome during interphase and mitosis;

- identify likely components to pass throught the nuclear pore;

- describe the principle features of RNA synthesis and processing to form mRNA;

- describe the structure of ribosomes and give an outline of the role of the nucleolus in their production.

The expression of genetic information: II - Protein synthesis and the genetic code

The expression of genetic information: II - Protein synthesis and the genetic code

7.1 Introduction

The events in the nucleus are now behind us. We have seen how it has been involved in the production of the various RNA molecules needed for protein synthesis but its role is now 'finished' and we move to the cytoplasm for the completion of the process. Proteins serve several roles. They may be enzymic, in which case they may be synthesised and released to move freely in the cytosol. Some enzymes and all membrane proteins are incorporated into membranes while others are manufactured for export to destinations outside the cell. The fundamental mechanism of protein synthesis is the same for all these processes and we will examine it first. This will be followed by a description of the various fates of the newly synthesised proteins.

7.2 Interpreting the genetic code

7.2.1 The sequence of nucleotides is a genetic code

translation

The factor which governs the specific nature of a protein is its sequence of amino acids whereas the information in mRNA is a sequence of nucleotides. The process by which the sequence of nucleotides in mRNA is converted into a sequence of amino acids in proteins is called translation.

Since there are only four nucleotides used in mRNA synthesis, if each nucleotide coded for a different amino acid there would only be four amino acids coded. If a sequence of two nucleotides coded for an amino acid this would give us 4x4 = 16 possible amino acids. We know that there are 20 different amino acids in proteins so a doublet code is not satisfactory. If we think of groups of three nucleotides working together this would give us 4x4x4 = 64 possibilities which is more than enough. Longer combinations eg quadruplets are possible but nature usually takes the simplest route and it turns out that the triplet code is the actual code. The sequence of these nucleotide which specifies each amino acid is called a codon.

SAQ 7.1

The enzyme RNase is a polypeptide containing 124 amino acid residues.

1) How many codons are there in the DNA which codes for this enzyme?

2) Assuming there is no overlap of codons what is the minimum number of nucleotides in this DNA?

7.2.2 tRNA molecules crack the code

amino acid
tRNA

There is at least one tRNA for each different amino acid and each tRNA is able to react with its own amino acid because of the action of aminoacyl tRNA synthetases. These enzymes catalyse the activation of amino acids and couple them to their specific tRNA molecules (Figure 7.1). There is a different synthetase for each amino acid.

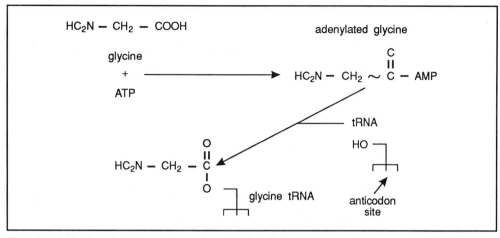

Figure 7.1 Activation of glycine and formation of glycine tRNA in a two step reaction catalysed by the aminoacyl synthetase specific for glycine. The curved bond in adenylated glycine indicates that it is a high energy bond. Note that the tRNA amino acid link is also a high energy bond ie the amino acid is still activated. The structure of tRNA shown in this figure is highly stylised. For a more detailed description of tRNA see BIOTOL text 'The Molecular Fabric of Cells'.

anticodon

peptide bond

The tRNA molecules act as the final adapter in this sequence because part of their structure, a three base group called the anticodon, recognises a complementary codon on the mRNA molecule and binds to it by hydrogen bonds. In this way the codon sequence in a mRNA molecule is translated into an amino acid sequence. The basic reaction in the synthesis of a protein is the formation of a peptide bond between the carboxyl group of a growing polypeptide chain and a free amino group on an amino acid. This process occurs with the tRNA molecule bound to its specific codon on the mRNA but the process occurs in a step-wise fashion rather than in the manner of a zipper. The process is shown diagrammatically in Figure 7.2.

The peptide bond is formed using the energy of the activated aminoacyl tRNA. Each amino acid added to the chain carries the energy which will be used for the addition of the next amino acid. If you look carefully at Figure 7.2 you will see that the codon sequence of mRNA is read two at a time. An alternative to this would be to line up all of the appropriate aminoacyl tRNA molecules on the mRNA and then to zip them up into a protein.

∏ Can you think why attempting to line up all of the appropriate aminoacyl tRNAs might be an unsuitable way to make a protein?

One disadvantage would be that a very large number of tRNA molecules would be required and they would be sitting around doing nothing until all the codons on the mRNA were filled. This would not be a very efficient use of tRNA molecules. The process would probably be slow.

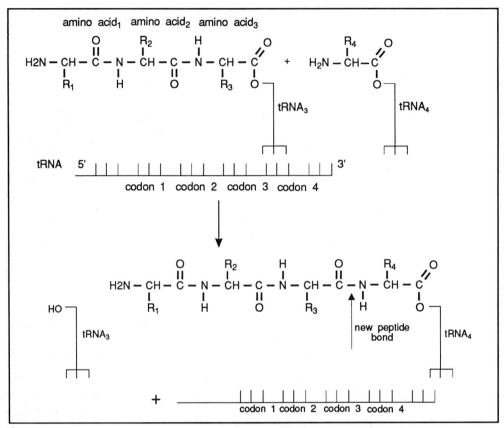

Figure 7.2 Synthesis of a protein is a step wise process. Two aminoacyl tRNA molecules bind to mRNA adjacent to each other. A peptide bond forms between the amino acids and the tRNA closest to the 5' end of the mRNA (by convention the left hand tRNA) is released.

7.2.3 Protein synthesis occurs on ribosomes

The question posed at the end of the previous section is a hypothetical one. This is because protein synthesis occurs on ribosomes and they are not large enough to house all the tRNA molecules which would be needed to do the job as a one step process. Furthermore, although the ribosome does not control the nature of the protein synthesised it does have a number of specific properties. Although prokaryotic and eukaryotic ribosomes differ from each other in a number of ways, they are both made of large and small subunits. They also have two grooves, one which accommodates the growing polypeptide chain and another for the mRNA molecule. The mRNA groove is long enough for about 35 nucleotides. Since mRNA molecules often contain 1500 nucleotides a lot of mRNA protrudes on either side of the ribosome. In fact there are only two tRNA molecules bound to the ribosome at any one time, the two binding sites being called the P(for peptide) and the A (for amino acid) (Figure 7.3).

A and P sites

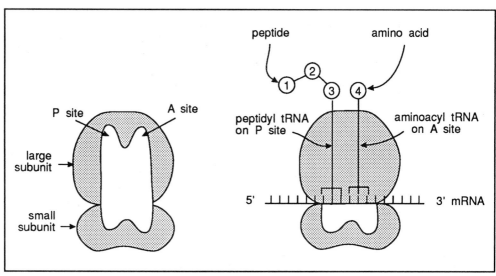

Figure 7.3 Structure and role of A and P sites on ribosomes.

These sites have the property that in order for a tRNA molecule to fit it's anticodon must be complementary to the codon on the mRNA. The P and A sites are situated such that tRNA molecules fit onto adjacent codons without leaving any gaps by way of 'spare' nucleotides. As shown in Figure 7.4 the process can be seen to be three parts.

the steps in protein synthesis

energy demand for protein synthesis

Starting with a ribosome in which the P (peptide) site is occupied the first step is the approach and binding of an aminoacyl tRNA to the vacant A (amino acid) site. In step two, the carboxyl end of the growing peptide is detached from the P site tRNA and forms a peptide bond with the amino end of the amino acid on the A site tRNA. In step three the 'naked' tRNA vacates the P site and the peptide tRNA on the A site moves to occupy this vacant site. However it is the ribosome that moves, not the tRNA. Thus the peptidyl tRNA stays bound to its anticodon but the movement of the ribosome by three nucleotides to the right transfers the peptidyl tRNA from the A to the P site. This movement of the ribosome is an energy requiring step and involves changes in conformation in one of the ribosomal proteins. When step three is complete the A site is vacant and the process can start again. Each cycle of three steps takes only a twentieth of a second and bacteria can synthesise a protein containing 400 amino acids in about 20 seconds. The cost of this, in terms of energy consumption, however, is quite considerable and protein synthesis consumes more energy than any other biosynthetic process. This is because a total of four high energy bonds are consumed to make each new peptide bond. One is used to form the activated amino acid linked to AMP and another to transfer this to its tRNA. Binding of the aminoacyl tRNA to the A site consumes a third and, as we have just seen, movement of the ribosome from the A to the P site uses the fourth.

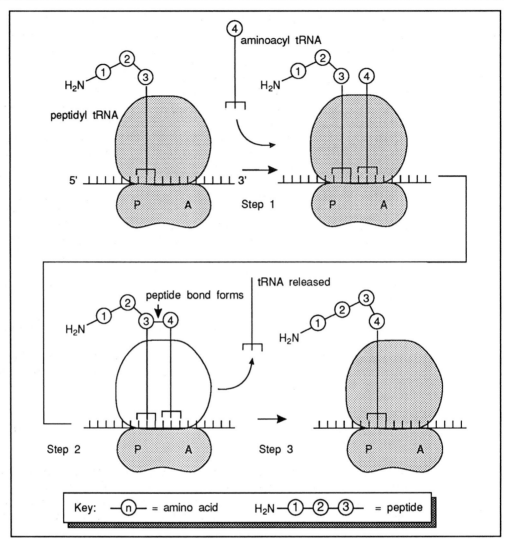

Figure 7.4 The three step process for the formation of a peptide bond, the details of which are described in the text. P = peptide site, A = amino acid site.

∏ By examining Figures 7.2 and 7.4, find out in which direction (3' 5' or 5' 3') mRNA is read. Is each protein made from its amino or its carboxylic acid end?

You should have come to the conclusion that mRNA is read in the dirction 5' 3' and that protein is made from its amino end. Note that each mRNA can be read several times to make many protein molecules.

SAQ 7.2

The side chain of the amino acid cysteine, which is -CH₂SH, can easily be chemically converted to -CH₃, which is the side chain of alanine (ie we can change cysteine into alanine). This conversion can be brought about even when cysteine is attached to its specific tRNA. If such modified amino acyl tRNA molecules are used in protein synthesis what effect do you think they would have on the amino acid sequence of the protein?

7.2.4 Protein synthesis involves initiation, elongation and termination

initiation factors

AUG

cap

ribosome
recognition

elongation

release factor

termination

Fully formed ribosomes, ie those containing large and small subunits, are not involved in initiation of protein synthesis. Rather, initiation involves the addition of components such as the large and small subunits, in a strict sequence (Figure 7.5). The initial event involves a special tRNA, initiator tRNA, which recognises the start code of protein synthesis, the codon AUG. AUG stands for the nucleotide sequence 5'-adenosine uridine guanosine 3'. This initiator tRNA is specific for start AUG codons and carries a methionine residue. The small ribosome subunit binds to a tRNA-met in a reaction requiring initiation factors, but the process is not well understood. The tRNA-met-small ribosome subunit-initiation factor complex now associates with a mRNA molecule using the 5' capped end as a recognition sequence. Binding now occurs between the initiator tRNA-met and the start codon AUG and when this has happened the initiation factors are released. Only now does the large ribosomal unit come into the picture. It binds, forming the complete ribosome, in such a way that the tRNA-met is on the P site. The A site is now ready to accept an aminoacyl tRNA molecule and protein synthesis is underway. Prokaryote initiation differs in that the methionine on the initiator tRNA-met is chemically modified to N-formyl methionine (ie it has a formyl group added to the amino group). Thus all proteins are synthesised with an N terminal methionine (N-formyl methionine in prokaryotes) but this residue is usually removed shortly after the protein is formed.

As we have seen above the codon AUG codes for the amino acid methionine and there maybe several such codons in mRNA, ie a protein might have the structure: met $aa_2.aa_3.aa_4.met.aa_6.aa_7.aa_8$... Thus in the messenger for that protein there will be AUG codons at the beginning of the messenger and at the 5th codon. Only one of these AUG codons is used as a start signal.

∏   Can you suggest a way in which the ribosome knows which AUG to use as the start signal?

We are limited here only by our imagination. Experiments show that removal of the cap on the 5' end of a mRNA molecule prevents ribosome binding and therefore translation; which is why the cap is referred to as the ribosome recognition sequence. It is possible, therefore, that the AUG codon used as the start signal is the one nearest to the 5' end of mRNA.

Protein synthesis now proceeds as described above, the formation of the peptide bond being catalysed by the enzyme peptidyl transferase. The overall process of adding amino acids to the growing peptide chain is called elongation and it proceeds until a stop codon in the mRNA is in the A site of the ribosome. A protein release factor then binds to the stop codon and causes peptidyl transferase to catalyse the addition of a water molecule (ie the peptide -tRNA link is hydrolysed) instead of an amino acid to the growing polypeptide (Figure 7.6). This severs the connection of the tRNA with the polypeptide which now moves away, spontaneously coiling into its secondary and tertiary structure. Ribosome disassembly now occurs and the small and large subunits, the final tRNA and release factor all separate. The events just described are referred to as termination.

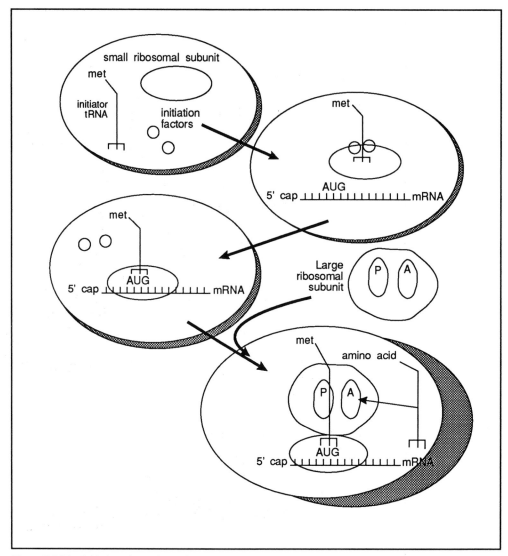

Figure 7.5 The initiation of protein synthesis.

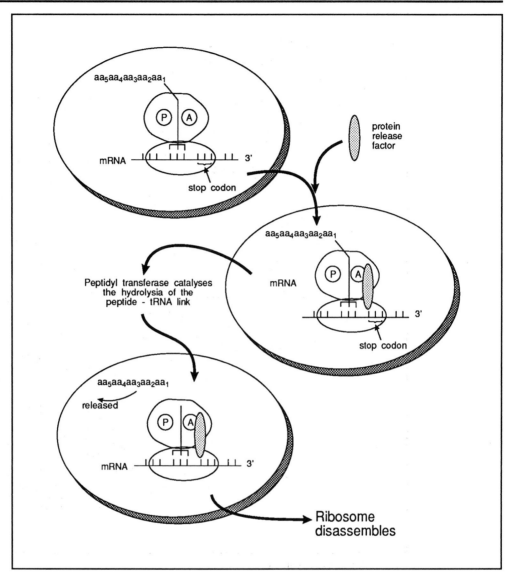

Figure 7.6 Termination of protein synthesis.

The major natural control point in the process of protein synthesis appears to be the activity of the initiation factor. Evidence suggests that it can be phosphorylated (ie phosphate groups added to it), by a protein kinase, which reduces its activity as an initiation factor. Note that protein kinases are enzymes that phosphorylate proteins using ATP as the phosphate donor. In this case the degree of phosphorylation of the initiation factor is crucial in the control of protein synthesis.

SAQ 7.3	Regarding protein synthesis which of the following apply to 1) initiation, 2) elongation, 3) termination? a) Requires peptidyl transferase activity. b) Requires recognition of a specific codon. c) Recognition of the capped 5′ end of mRNA. d) Requires a conformational change in a ribonucleoprotein. e) Involves tRNA-met. f) Involves binding of tRNA-met to the small ribosome subunit. g) Involves a release factor. h) Involves peptide bond formation. i) Involves formation of a free carboxyl group on an amino acid.

7.2.5 Binding of ribosomes to a mRNA forms polysomes

As we have seen above translation of the information in mRNA begins at the 5′ end and proceeds to the 3′ end. As soon as there is enough space free on the 5′ end of the mRNA another ribosome can bind and begin to synthesise protein. This results in there being numerous ribosomes, spaced as close as 80 nucleotides apart, along the mRNA molecule. This combination of mRNA and ribosomes is referred to as a polyribosome or polysome. Because of the association of several ribosomes with mRNA in a polysome its molecular mass is very much increased.

polysomes

∏ Can you suggest a way to separate polysomes from single ribosomes?

Clearly use could be made of the mass of the polysome complex. A very good way is by density gradient centrifugation. The identification of a particular species of mRNA from polysomes provides very good evidence that the corresponding DNA coding sequence is being transcribed and translated. Whereas eukaryotes read their mRNA as described above and produce a single protein from a mRNA molecule, this is not the case with prokaryotes. Here there may be several start and stop signals present in a single (polycistronic) mRNA but instead of the ribosome falling apart at a stop signal it releases the protein already made, glides over the stop signal to the next start signal and begins to make the next protein. Eukaryotes often cleave newly-synthesised proteins into a number of smaller ones, which have different enzyme properties. In addition they also produce enzymes which are multifunctional. Thus they appear to achieve the same ends as prokaryotes but by different means.

polycistronic messenger

7.2.6 Protein turnover

We have just examined how protein synthesis is brought about in cells. In addition to being synthesised, a protein can also be degraded and its concentration at any one time depends upon the balance between its rate of synthesis and rate of degradation. The synthesis and degradation of proteins is referred to as protein turnover. The life span

protein turnover

for an average protein in animals is about two days but the range varies from minutes to months. The most short-lived proteins are enzymes which catalyse rate-limiting steps in metabolic reactions. The levels of these proteins are found to vary considerably with environmental conditions. The only way to achieve this is to have a system for their rapid degradation. Their concentration can then be accurately controlled by fine-tuning the rate of their biosynthesis. This probably appears to be very wasteful, but it is a necessary evil if metabolism is to be operated in a controlled way.

7.2.7 Many inhibitors have helped the understanding of protein synthesis

As you can see from the above, protein synthesis is a complex process and it has taken many years to reach our present level of understanding. Of some considerable use in the elucidation of this process were compounds which inhibit the process in a specific position. The application of such compounds effectively breaks the whole process into a number of shorter sequences and this makes it easier to sort them out. Table 7.1 shows some of these inhibitors.

Tetracycline	blocks binding of aminoacyl-tRNA to A site of ribosome
Streptomycin	prevents the transition from intitiation to elongation of protein synthesis
Chloramphenicol	blocks peptidyl transferase reaction on ribosomes
Rifamycin	binds to RNA polymerase and blocks RNA synthesis
all of the above are active only on prokaryotes	
Cyclohexamide	inhibits peptidyl transferase reaction on ribosomes in eukaryotes
Amanitin	binds to RNA polymerase II and inhibits RNA synthesis in eukaryotes

Table 7.1 Action of inhibitors of protein synthesis.

7.2.8 Mutations involve changes in the genetic code

Sickle cell anaemia refers to a condition in which the red blood cells are sickle-shaped rather than their usual round flattened shape. People who have this disease cannot transport oxygen very well in their blood and this has been shown to be due to their haemoglobin (Hb) being abnormal. Abnormal Hb in this condition differs from normal Hb in only one of its 287 amino acid residues, a glutamic acid residue being replaced by a valine residue. Thus the gene coding for the abnormal Hb differs from that coding for normal Hb in only one codon. This change in genetic information is referred to as a mutation. The mutation referred to here caused the codon for glutamic acid to be changed to that of valine. Thus a protein was still produced despite the mutation.

mutation

SAQ 7.4	What would happen to the synthesis of a protein if a mutation occurred such that the mRNA contained a non-coding sequence?

genotype   The sum total of the information contained in an organism's DNA is referred to as its
             genotype. This refers to its genetic constitution. The expression of genetic information
             is affected by a number of factors, including the environment in which the organism
             lives. Not all of the genetic information is expressed all of the time so cells with the same
             genotype might be quite different in different environments. We have generated a term
phenotype   to specify the appearance of the organism, or part of it. This is known as its phenotype.

Thus the muscle cells and kidney cells of a single animal have the same genotype but
are phenotypically quite different. We shall learn more of genotype and the generation
of different phenotypes in the chapter dedicated at looking at cell differentiation
(specialisation).

7.3 The genetic code

7.3.1 Cell-free systems allow the elucidation of the genetic code.

Cell-free systems are made by breaking cells open under conditions in which
biochemical activity is retained. Carefully prepared cell-free systems of prokaryotes are
able to synthesise proteins especially if amino acids and ATP are added to supplement
those already present. A very exciting discovery was reported by two scientists, M.
Niremberg and H. Matthaei, in 1961. They synthesised an artificial RNA molecule
containing only uracil nucleotides (poly-U). When added to a cell-free bacterial system
they discovered that a protein was made which contained only phenylalanine.

| SAQ 7.5 | What does the above observation suggest about the genetic code and poly-U? |

This discovery catalysed enormous activity in this field. A wide variety of synthetic
mRNA molecules were subsequently used but of particular importance was the
discovery that a mRNA molecule consisting only of a single trinucleotide could be used
to direct amino acid incorporation in cell-free systems. If such a specific trinucleotide
was incubated with a cell-free system, ribosomes could be subsequently isolated which
contained the trinucleotide, a tRNA molecule and an amino acid. The amino acid was
specific to the codon coded by the trinucleotide. In this way codons for all the amino
acids could be found.

| SAQ 7.6 | What other problem did the use of trinucleotides solve in the quest for the genetic code? (Hint; think what would happen if all the spaces between the words in this sentence were removed). |

7.3.2 The code

The genetic code is listed in Figure 7.7 All 64 combinations of 3 nucleotides (ie codons)
are listed.

Second nucleotide base			
U	**C**	**A**	**G**

First nucleotide base	Second nucleotide base				Third nucleotide base
U	UUU ⎱ phe UUC ⎰ UUA ⎱ leu UUG ⎰	UCU ⎱ UCC ⎰ ser UCA ⎱ UCG ⎰	UAU ⎱ tyr UAC ⎰ UAA c.t.* UAG c.t.*	UGU ⎱ cys UGC ⎰ UGA c.t.* UGG trp	U C A G
C	CUU ⎱ CUC ⎰ leu CUA ⎱ CUG ⎰	CCU ⎱ CCC ⎰ pro CCA ⎱ CCG ⎰	CAU ⎱ his CAC ⎰ CAA ⎱ gln CAG ⎰	CGU ⎱ CGC ⎰ arg CGA ⎱ CGG ⎰	U C A G
A	AUU ⎱ ile AUC ⎰ AUA ⎱ AUG met	ACU ⎱ ACC ⎰ thr ACA ⎱ ACG ⎰	AAU ⎱ asn AAC ⎰ AAA ⎱ lys AAG ⎰	AGU ⎱ ser AGC ⎰ AGA ⎱ arg AGG ⎰	U C A G
G	GUU ⎱ GUC ⎰ val GUA ⎱ GUG ⎰	GCU ⎱ GCC ⎰ ala GCA ⎱ GCG ⎰	GAU ⎱ asp GAC ⎰ GAA ⎱ glu GAG ⎰	GGU ⎱ GGC ⎰ gly GGA ⎱ GGG ⎰	U C A G

Figure 7.7 The genetic code is the sequence of nucleotides in a codon which specifies a given amino acid. The code is universal ie the code is the same in bacteria, plants and animals. Recently some exceptions have been described but these are extremely rare and will not be discussed further here. Note that c.t.* are stop or chain terminating sequences (The abbreviations used for the amino acids are listed in the appendix).

∏ Examine Figure 7.7 and decide whether: 1) more than one codon has the same meaning and 2) one codon codes for more than one amino acid.

The answer to 1) is yes. For example GUU, GUC, GUA and GUG all code for valine. The answer to 2) is no. Indeed only methionine and tryptophan have less than two codons.

degenerate code

Because more than one codon can determine the same amino acid the code is said to be degenerate.

∏ Can you think of any advantage in having a degenerate code?

We have seen earlier that the mRNA for abnormal Hb in sickle cell anaemia differs from the mRNA by only one codon. This is common in mutation. So, if only 23 codons specified amino acids, the rest of them being nonsense, a mutation (a change in a single nucleotide) in any one of the 23 would be much more likely to produce a nonsense codon.

Strange as it may sound there is a pattern to the degeneracy of the code. Where degeneracy occurs it tends to occur in the third nucleotide of the codon with U and C and also A and G being equivalent to each other. In eight cases, ie the codons for serine, leucine, proline, arginine, threonine, valine, alanine and glycine all four final nucleotides are equivalent to each other (eg serine can be coded for by UCU, UCC, UCA or UCG). This has the advantage, already described above, that a mutation in the third nucleotide has a lower chance of significantly modifying the code than a change in the second and certainly the first. We have seen that no codon codes for more than one amino acid ie the code is not ambiguous.

SAQ 7.7

Why would an ambiguous code be detrimental?

SAQ 7.8

You are provided with a synthetic mRNA of sequence 5' UUU UCG CAU UGG 3'. Using Figure 7.7 decide:

1) what amino acid sequence is encoded by this sequence;

2) what the DNA sequence was from which the mRNA was produced.

stop signals
start signals

Note also in Figure 7.7 that there are three codons which signify the stop signal. These are UAA, UAG and UGA. We have seen already that the code AUG, which codes for methionine, is the starting sequence and there is only one of these.

∏ Here is a nucleotide sequence, can you spot a part of this nucleotide sequence that codes for a short peptide? UUUAGUGGUAUGGCCAUCCUCUGACCCAUU.

If you look along the sequence, you will come to the triplet AUG. This could act as the start codon. A little further along is the sequence UGA, a stop codon. We will dissect this sequence out: AUGGCCAUCCUCUGA.

This would code for the amino acid sequence: Met,ala,ile,leu (ie a tetrapeptide).

Can you see that the start codon (AUG) tells us where to start reading the messenger RNA, rather like the way that a capital letter tells us where to start reading a sentence? The stop codon acts rather like a full stop in a piece of text. It tells us where a particular messenger ends.

7.4 Post translation events

7.4.1 Proteins do not always stay where they are made

If we consider proteins as a whole we can say that they have a number of fates. Many proteins are enzymes and function in the cytosol. Many others are required by chloroplasts or mitochondria and these are transported across the membranes surrounding their respective organelles. All cells produce proteins that will become part of a membrane and finally some cells produce proteins which are destined for secretion out of the cell. We will now examine how these various fates are achieved.

7.4.2 The endoplasmic reticulum and Golgi apparatus function in protein secretion

We learnt of some details of the endoplasmic reticulum and the Golgi body in Chapter 2. Here we remind you that the endoplasmic reticulum (ER) is thought to form a single continuous convoluted sheet of membrane enclosing a single space. This space often occupies as much as 10% of the cell's volume and is continuous with the space within the nuclear envelope. The ER is very often associated with the Golgi apparatus, which consists of membrane-bound cisternae or dictyosomes forming a series of stacks. The

Golgi cisternae

individual Golgi cisternae are separate entities and unlike the ER do not form a single continuous membrane system. There are usually numerous vesicles associated with these two organelles (Figure 7.8).

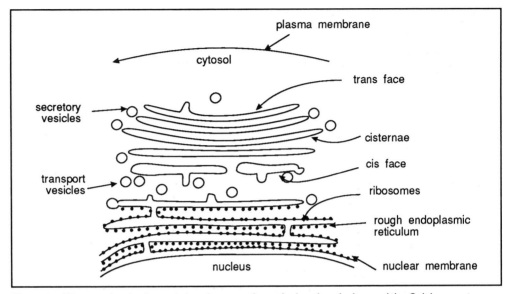

Figure 7.8 Diagram showing interrelations between the endoplasmic reticulum and the Golgi apparatus.

As is usual in biology, progress in the study of protein secretion was helped by the use of a cell type which specialises in this process.

acinar cell Such cells are the acinar cells in the pancreas, whose function is to synthesise and secrete a number of digestive enzymes, including RNase, DNase, amylase and lipase. Most of the proteins synthesised in the acinar cells are secreted into the alimentary tract and it is possible, to trace the intracellular pathway followed by these proteins. But how can this be achieved?

One technique would be to produce radioactive proteins by incubating the cells with radioactive amino acids and then to follow where the radioactivity goes by using autoradiography of these sections. These autoradiographs and sections are viewed and compared at high magnification. Figure 7.9 shows the result of such an experiment.

Experiments such as those described in Figure 7.9 can be extended by extracting the soluble components and demonstrating that the radioactive material is protein. This experiment, therefore, suggests that proteins are synthesised on rough ER, passed in transport vesicles to the Golgi apparatus and packaged again into secretory vesicles ready for secretion out of the cell. It is implicit in this sequence that the protein passes through the ER membrane as it is being synthesised.

SAQ 7.9

Describe the appearance of the autoradiograph in Figure 7.9 if radioactive amino acids were fed to the tissue for only a short time, to be followed by exposure to non-radioactive amino acids. For this, sections of the tissue would be taken at time intervals and exposed to photographic films. After development the films would be viewed at high magnification and compared with the tissue section.

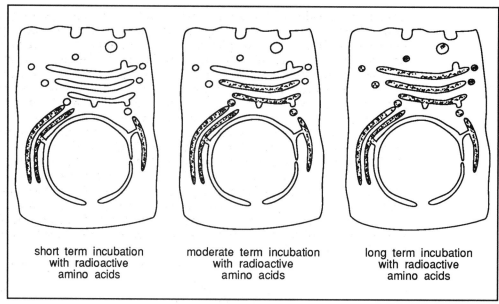

short term incubation
with radioactive
amino acids

moderate term incubation
with radioactive
amino acids

long term incubation
with radioactive
amino acids

Figure 7.9 Diagrammatic illustration of results obtained when acinar cells are incubated with radiolabelled amino acids and sections prepared for autoradiography. The shaded areas indicate where radioactive material is found. Note that radioactivity is detected first over the ER, then over the ER plus the Golgi apparatus and finally over the ER, Golgi apparatus and secretory vesicles.

7.4.3 Ribosomes bind to specific proteins in the ER

ribophorins

Two specific glycoproteins function to bind ribosomes to the ER. These proteins, which span the ER membrane, are called ribophorins and they are found exclusively embedded on the rough ER. A mechanism, the detail of which is unknown at present, prevents the migration of ribophorins from the rough to the smooth ER. Ribophorins and their restraining mechanism are considered to account for the flattened structure of the rough ER in comparison with the circular outline of smooth ER. The larger ribosomal sub-units are bound by the ribophorin. Ribophorin binding is not the only way in which ribosomes are attached to the ER. The discovery of the second method of attachment used the fact that rough ER can be separated from other cellular membranes by differential centrifugation. During the extraction process, the rough ER forms vesicles with ribosomes attached to their outer surface, the vesicles being called

microsomes

microsomes. Protein synthesis can occur both on the ribosomes attached to microsomes and also on free ribosomes but the product is not the same. When a single species of mRNA, coding for a secreted protein, is used for protein synthesis the product from free ribosomes has an extra polypeptide zone at the N terminal end containing about 20 amino acid residues, as compared with the protein produced from the same mRNA

leader
sequence

attached to the ER ribosomes. The 20 amino acid portion is called the leader sequence. The evidence suggests that this sequence is in fact a signal sequence which is recognised

signal
sequence

by a site on the rough ER. This binding site is considered to provide a channel through which the growing peptide moves as it is synthesised and is the means by which the vectorial (directional) aspect of the process is achieved. We may call this binding site the

ER channel
complex

ER channel complex. Corroborative evidence for the significance of the signal sequence comes from the fact that other secretory proteins similarly have a leader sequence when synthesised on free ribosomes but not when synthesised on rough ER.

∏ Can you think what might happen to the signal sequence as the protein passes through the ER membrane?

Careful examination of microsomes show that they contain a specific protease, on their lumen (inside) side, which catalyses the removal of the signal sequence. Further examination showed that the picture was still incomplete. The signal sequence alone does not appear to recognise the ER channel complex. Recent evidence has shown the

signal recognition protein

presence in the cytosol of a protein that binds to the signal sequence and also binds to the ER channel complex. This signal-recognition-protein has a crucial role because further translation of the mRNA is prevented until the bound signal-recognition-protein binds to the ER channel complex. The exact nature of the ER channel complex is not known but it and the ribophorins are the means by which the protein synthetic machinery is attached to the ER. One final point should be made here. The N terminal signal sequence of amino acids described above is not found on proteins whose fate it is to stay in the cytosol. All proteins which combine in some way with the ER and other membrane-bound organelles, such as chloroplasts and mitochondria, contain N-terminal signal sequences. Thus these appear to be a common 'address' system. We will return to the chloroplast and mitochondria situation after completing the description of secretion. Now examine Figure 7.10 and read through this section again before attempting to answer SAQ 7.10.

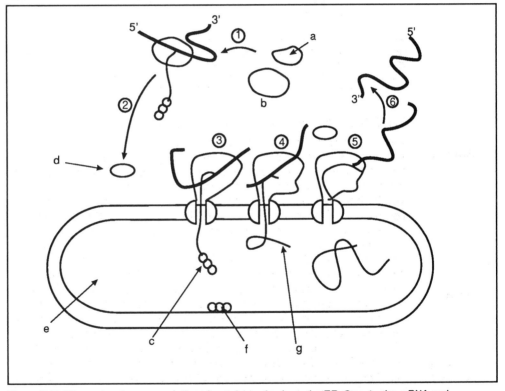

Figure 7.10 Diagrammatic representation of protein synthesis on the ER. Step 1 - the mRNA and ribosomal subunits form complexes and protein synthesis begins. The signal sequence is made. Step 2 - the signal sequence binds with the ER channel complex and ribosomes attach to ER. This process also involves another protein - a signal recognition protein. Step 3 - the singal sequence moves into the ER lumen and is removed (Step 4). Step 5 - the rest of the protein is synthesised and released into the ER lumen. Step 6 - the ribosomal subunits and mRNA dissociate.

SAQ 7.10	1) Using the appropriate words from the list below, and the text to help you label the parts a-g in Figure 7.10. Removed leader sequence; ER lumen; amino end of growing peptide; small ribosomal subunit, large ribosomal subunit, leader sequence, membrane accepter protein. 2) Explain where and how ribophorin participates in this sequence of events.

7.4.4 The addition of carbohydrate residues occurs in the ER

glycoprotein

glycosylation

glycosyl
transferase

Proteins synthesised on the ER and passed into its lumen are glycoproteins. This means they contain covalently attached carbohydrates. Proteins made on ribosomes not attached to the ER do not contain carbohydrates. The process of adding the carbohydrates is called glycosylation and it occurs in the lumen of the ER. The carbohydrate added to proteins is predominantly one species of oligasaccharide containing 14 monosaccharide residues. The specific sugars in it are of only three types; N-acetylglucosamine, mannose and glucose. The oligosaccharide is attached mainly to the -NH_2 group on the side chain of asparagine (asn) residues which are found in the target sequences asn-X-serine or asn-X-threonine, where X is any amino acid. Thus only a small number of asn residues are glycosylated. The addition of the oligosaccharide is catalysed by an enzyme (glycosyl transferase) embedded in the ER with its active site exposed on the lumen side. The addition appears to occur while the protein is still being synthesised (ie before the protein is released into the ER lumen). This may imply some sort of spatial relationship between the ER channel complex protein and the glycosyl transferase.

SAQ 7.11	1) What problem might arise if glycosylation occurred on the cytosol side of the ER membrane? 2) Similarly what would be the consequence if every asn residue was glycosylated instead of simply those in the target sequences? 3) Finally which amino acids does the glycosyl transferase enzyme recognise?

dolichol

The glycosyl portion of the glycoprotein begins life as a glycolipid for it is on a membrane-bound lipid that the oligosaccharide is built up. This lipid, named dolichol, protrudes into the ER lumen and it is prepared for action by being phosphorylated by reaction with ATP. The sugar residues in the form of sugar nucleotides are produced in the cytosol and one by one the 14 sugar residues are added to dolichol. Enzymes in the ER lumen catalyse this addition to dolichol forming a branched oligosaccharide, the link with dolichol being a pyrophosphate moiety. The pyrophosphate bond is a high energy bond thus activating the oligosaccharide and the glycosyl transferase uses this energy in catalysing its addition to the protein. We have represented this process in Figure 7.11.

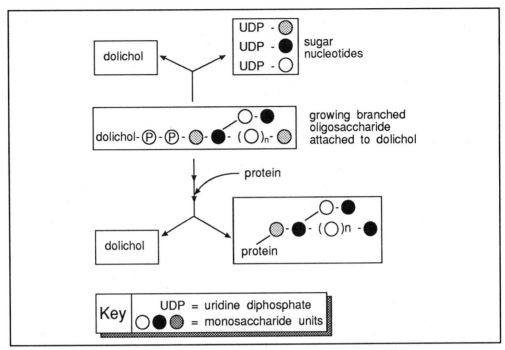

Figure 7.11 The synthesis and addition of oligosaccharides to proteins in the ER lumen.

7.5 Protein secretion

7.5.1 The Golgi apparatus is the delivery system of secreted proteins

transport
vesicles

Large molecules synthesised in the ER are packaged into small transport vesicles at specific places in the ER. Vesicles containing proteins destined for secretion fuse with the Golgi apparatus where the glycosyl portions of the proteins undergo chemical modification.

changes in
sugar residues

The modifications are catalysed by enzymes embedded in the Golgi membrane with their active sites exposed on the lumen side. The glycosyl portions, which began containing 14 sugar residues finishes with 14 but they now contain galactose and N-acetylneuraminic acid residues but no glucose residues. The exact details do not concern us but the point to be realised is that specific modifications occur which are brought about by the Golgi apparatus.

cis/trans faces
of the Golgi

The Golgi apparatus consists of a 'cis' or forming face and a 'trans' or maturing face (see Figure 7.8). The cis face is usually close to the ER, the trans face is towards the plasma membrane. Transport vesicles from the ER are thought to fuse with the cis face. Secretory vesicles containing the modified proteins are produced on the trans face and fuse with the plasma membrane, releasing their contents into the extracellular space. The transfer of material in the Golgi lumen from the cis face, through the stacks of cisternase, to the trans face is complex because, as mentioned above, they are not in lumenal contact. The details of this process are not known, but a suggested scheme is illustrated in Figure 7.12 and described below.

Secretory vesicles produced from the trans face fuse with the plasma membrane and release their contents. The membrane of the vesicle becomes part of the plasma membrane but it is recovered by endocytosis forming a recovery vesicle. Multiples of these fuse with transport vesicles from the ER and form the cis face of the Golgi apparatus. As proteins 'move' through the Golgi apparatus they are modified by the enzymes on the Golgi membranes. Although the scheme is speculative it is conservative of the Golgi enzymes, the recovery vesicle bringing them back into functional use.

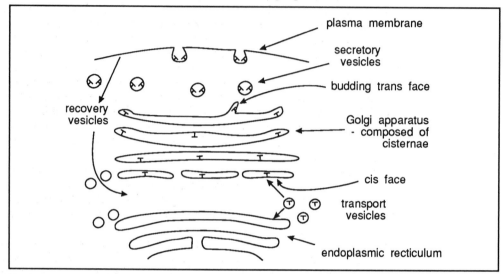

Figure 7.12 Possible mechanism for operation of the Golgi apparatus. See the text for details.

To complete the picture of the Golgi apparatus it is important to realise that it has some additional functions. One of its major uses in plant cells is to manufacture many of the raw materials for cell wall production and to secrete these into the wall compartment where they are polymerised to form the mature wall.

| **SAQ 7.12** | Which of the following is true (T) or false (F) regarding the secretion of protein. |

1) Secreted proteins are incorporated into ER transport vesicles.

2) Sugar residues are removed.

3) Sugar residues are added.

4) Binding to Golgi apparatus membrane is required.

5) Budding of vesicles occurs at the cis face of the Golgi apparatus.

7.6 Membrane synthesis

7.6.1 The ER is also involved in membrane synthesis

The synthesis of membranes involves the production of new phospholipids and proteins. Phospholipids are generated on the cytosol face of the ER using raw materials generated in the cytosol. This increases the area of the bilayer on the cytosol side but the balance is redressed by the transfer of phospholipids from the outer to the inner half in a reaction called flip flop. Although this phenomenon has been demonstrated, so far no enzyme has been isolated which catalyses it. The protein part of the membrane is considered to arise in two ways. There is evidence that certain protein species, released into the ER lumen as if for export, become part of the membrane by binding to the ER lumen face. This is considered to occur by the protein carrying a second signal sequence which is recognised by a receptor in the ER membrane. Alternatively a cytosolic protein may be inserted from the cytosol side by a similar mechanism. A third possibility has been studied in cells which bind large quantities of a single protein species in a particular membrane. An example is the vesicular stomatitis virus. When this virus infects a cell it causes the ER to produce an envelope around the virus and this envelope contains only one membrane protein, the G protein. Studies show that G protein is produced as if for export but it is not released into the ER lumen. Instead the protein stays anchored in the membrane and becomes an integral membrane protein. We do not know exactly how this occurs. It is possible that there is a specific signal sequence on proteins which are destined to become part of the membrane. By one or more of the above mechanisms new proteins become embedded in the ER. This, coupled with the production of new phospholipids, allows new ER membranes to be generated. These can now be incorporated into their target zones, via the production of vesicles by exocytosis and fusion of the vesicles with the target zone by endocytosis.

flip flop

vesicular stomatitis virus

7.6.2 How do chloroplasts and mitochondria expand their membranes?

These two organelles do not increase their membrane area by the incorporation of vesicles. Evidence suggests that the phospholipids found in mitochondria are generated on the smooth ER and transported to the mitochondria by carrier proteins called phospholipid transfer proteins. Each transfer protein recognises a single phospholipid species and can withdraw it from the ER membrane on which it has been made.

phospholipid transfer proteins

Π What major problem is faced by the phospholipid transfer protein in transporting phospholipid from the ER to the mitochondria? (Hint, think about the solubility of phospholipids in water).

The phospholipid is not very soluble in water, but it has to be transported across the aqueous cytosol.

The problem is solved by the protein wrapping itself around the phospholipid so that it is not exposed to the cytosol. The bound phospholipid is released when a suitable membrane is encountered. We can postulate that the flip flop transferase catalyses the transfer of phospholipid from the outer side of the bilayer to the other.

flipflop transferase

Before completing this story we should examine the chloroplast. Chloroplasts synthesise their own phospholipid by a process similar to that of ER. The evidence suggests that it occurs on the stroma side of the inner membrane and also require the operation of the flip flop transferase. In both chloroplasts and mitochondria a problem arises over how phospholipid gets into the other membrane ie into the inner mitochondrial and the outer chloroplast membrane. (Remember that both mitochondria and chloroplasts are surrounded by a double membrane).

Before discussing the answer to this problem these are other aspects of it which need to be described. Think of the incorporation of proteins into mitochondrial and chloroplastic membranes. Let us consider the mitochondrion first (Figure 7.13).

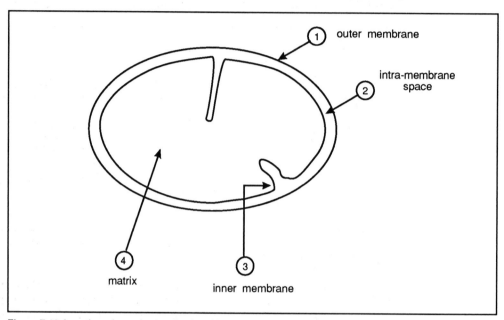

Figure 7.13 Location of proteins in mitochondria.

Proteins are found in all four locations indicated, some as integral membrane proteins and some dissolved in the fluid between membranes and in the matrix. Some of these proteins are coded by mitochondrial DNA and synthesised on mitochondrial ribosomes. Most, however, are coded by the nucleus, synthesised in the cytosol on polysomes and imported into the mitochondria. As noted some proteins are destined for membranes and will have large hydrophobic sections. Others will be enzymes required in the mitochondrial matrix.

7.6.3 Proteins can be imported by post translational import

co-translational and post-translational import

The movement of a protein into the ER lumen during protein synthesis is referred to as co-translational import because the translation and importational parts occur together. It is possible for proteins synthesised in the cytoplasm to be imported into the ER lumen and this is referred to as post-translational import (PTI). It is an energy-requiring process. In experiments studying PTI into the ER, the cytosolic protein was found to contain about 20 amino acid residues at the N terminal end not found in the final imported protein. The removal of these amino acid residues prior to the start of the experiment prevented its import into the ER.

Signal sequences have been found on all proteins imported by PTI and it appears to be an essential requirement. Unravelling of the protein's secondary and tertiary structure is also a requirement. It is difficult to imagine how a membrane-bound translocator can open up a channel large enough to allow a folded protein to go through. In certain **unwinding PTI** proteins unwinding can be prevented by binding with particular reagents. In the presence of these binding agents protein can not be imported by PTI. This suggests that unwinding is required for PTI. Figure 7.14 shows one way in which PTI may occur.

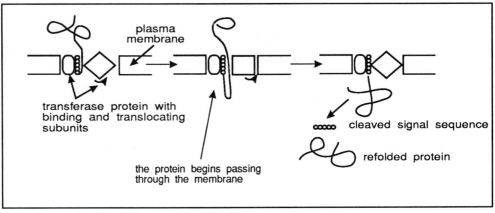

Figure 7.14 A proposed mechanism for post translational import of protein. Note that as the protein passes through the membrane, it unfolds.

The mechanism shown in Figure 7.14 would result in the release of the imported protein from the protein by dissociation from the signal sequence. A number of other possibilities exist which retain the protein in the membrane. In these processes the N terminal signal sequence is considered to act as start transfer signal which remains anchored to the membrane throughout translocation. A second type of signal sequence occurs, called the stop-transfer signal, which serves to stop translocation of the protein. Thus proteins with such stop translation signals would be retained in the membrane.

Many proteins synthesised in the cytoplasm become part of mitochondria or chloroplasts as membrane proteins and enzymes and this is considered to occur by PTI. However, the mechanism here must differ from that in Figure 7.14 because of the double membranes, unless we postulate energy-driven carriers operating at *both* membranes which is not out of the question. A phenomenon, occurring in bacteria which suggests movement through both membranes at *once*, provides a clue as to what might be happening in mitochondria and chloroplasts.

7.6.4 Contact sites may be involved in protein transport in mitochondria and chloroplasts

Gram negative bacteria produce proteins in the cytosol some of which become incorporated into the outer membrane of their cell walls. Thus they must have been transported through the inner (plasma membrane) to the outer (cell wall) membrane. This is analogous to the situation in mitochondria and chloroplasts. These bacteria are found to possess zones where their inner and outer membranes come together, called **contact sites** contact sites and experiments using antibodies to outer membrane proteins show that they appear first in the outer membrane near to a contact site. This is interpreted as suggesting that PTI occurs through the contact site. There is both biochemical and structural evidence for a similar phenomenon occurring in mitochondria and chloroplasts.

If cell-free fractions containing mitochondria and a protein known to be imported by PTI are cooled to 5°C the process stops. In many cases the imported protein is stuck part way into the mitochondrion. This can be demonstrated by the fact that its N terminal portion has already been removed by proteases in the mitochondrial matrix while the remainder of the protein can still be attacked by externally added proteases. Thus this protein appears to go through both membranes at once and, clearly, this would be a lot simpler if the inner and outer membrane were fused at that point. Such contact sites have been noted in electron micrographs of both mitochondria and chloroplasts. Experiments involving cooling of mitochondria reveal that insertion of the start transfer sequence uses the energy provided by the membrane potential but subsequent transfer of the rest of the molecule requires ATP. It is thought that ATP hydrolysis is involved in the catalysis of protein unfolding. You will remember that chloroplasts do not generate a membrane potential and here ATP hydrolysis drives both parts of the process. Figure 7.15 illustrates the structure of a contact site in a chloroplast.

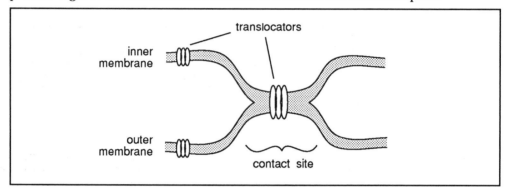

Figure 7.15 Illustration of a contact site. Two types of translocators are illustrated, one spans a single membrane, the other spans two at the contact site.

7.6.5 How do contact sites work?

translocator

We do not know the exact answer to this question but we have some ideas. Movement from the cytosol into the mitochondrion could be achieved using the mechanism described in Figure 7.14 assuming that the translocator spanned the contact site. This process releases the protein after cleavage of the start signal. This released protein could be an enzyme or it may be destined for insertion into the inner mitochondrial membrane. Evidence suggests that this is also the route used for proteins which become embedded in the cristae. Thus it appears that proteins are first translocated through contact sites into the lumen of the organelle where they may remain in solution or become incorporated into the inner face of the membrane.

We do not know if the contact sites are permanent or temporary or if they transport membrane proteins and soluble proteins by the same mechanism. You may remember we came into this discussion as a result of enquiring how phospholipid could move from the outer to the inner mitochondrial membrane and from inner to outer chloroplast membrane. It is considered possible that phospholipids can also migrate through contact sites. Although no details are known of this process it is possible that it could be catalysed by phospholipid transfer proteins located as peripheral proteins between inner and outer membranes (Figure 7.16).

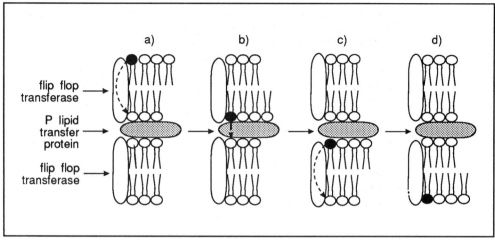

Figure 7.16 Scheme to explain movement of phospholipid from outer to inner membrane via a contact site. a) P lipid has been inserted into the outer face of the outer membrane and in b) has been transferred to the inner face by the flip flop transferase. From here a bound phospholipid transfer protein removes it and inserts it into the outer face of the inner membrane, c) from which, in d), it is transferred to the inner face of the inner membrane by another flip flop transferase. This type of scheme could operate in both mitochondria and chloroplasts.

SAQ 7.13

What organelle other than chloroplasts and mitochondria has a double membrane and how is it proposed that membrane protein and phospholipid move from its outer to its inner membrane?

In conclusion we would like to emphasise that, whereas the biosynthesis of a protein is a very striking event in its own right it is, in many cases, just the starting point of a journey. Proteins need to be distributed to particular sites either as naked proteins or packaged in a vesicle. 'Addresses' are needed in both cases and there are either leader sequences on proteins or signal recognition proteins in vesicle membranes. The correct reading of these addresses ensures the correct delivery of the protein. Much is known about this system but there is still much to learn.

We have used the last two chapters to examine the expression of genetic information in two chapters. We have seen how this genetic information is used to make particular proteins that are to be used in special parts of the cell's machinery. We have also learnt something of the mechanisms involved. How is this genetic information provided to each cell? The answer is by the process of cell division, which is the topic of the next chapter.

Summary and objectives

This chapter takes up the story of the expression of genetic information from the point at which the mRNA and ribosomes have been synthesised and released into the cytoplasm. We have examined the process of translating the nucleotide sequence of mRNA into the amino acid sequence of protein. We have learnt of the roles of tRNA and ribosomes in this process. We have also learnt that the genetic code is universal and unambiguous. We have spent quite a considerable time examining the fates of the proteins that are produced. We have seen that proteins destined to become enzymes in the cytosol are synthesised on free polysomes. The proteins which are to be secreted or must enter membrane-bound organelles are produced, carrying signal sequences, on the endoplasmic reticulum. We have followed the fate of such proteins including their glycosylation within the lumen of the endoplasmic reticulum and their subsequent packaging in vesicles. We have learnt of the role of the Golgi apparatus in post-translational protein processing.

Now that you have completed this chapter, you should be able to:

- describe in general terms, how a protein is synthesised from twenty different amino acids and assign roles to the four types of nucleic acid found;

- identify polysomes in electron micrographs;

- explain the meaning of the genetic code and solve problems of amino acid sequencing, using a table of codons;

- explain what mutations are and how they might cause changes in protein function;

- analyse and interpret presented data concerning protein synthesis and the genetic code;

- explain the meaning of autoradiography and deduce a sequence of events from data provided;

- describe the roles of the endoplasmic reticulum, the Golgi apparatus and signal sequences in the post-translational processing of proteins.

Cell growth and division

Cell growth and division

8.1 Introduction

The growth and development of every living organism depends on the growth and multiplication of its cells. Unicellular organisms reproduce by cell division, a single cell giving rise to two new daughter cells. Multicellular organisms, on the other hand, develop from a single cell, called the zygote, and it is the multiplication of this cell and its descendants that determines the development and growth of the individual. The requirement for cell growth and multiplication does not cease when an organism reaches maturity. There is still a continual demand for cell division in order to replace cells lost through natural wear and tear. In an adult human this represents the need to manufacture many millions of new cells every minute.

Cell growth and division require duplication of all of the cell's constituents so that, upon cell division, each daughter cell is fully functional. For most of the cell's components duplication need not be controlled exactly. If a particular molecule or organelle is present in multiple copies, then it is sufficient for the cell to ensure that the number of copies roughly doubles before cell division and that each daughter cell inherits approximately half. However, an absolutely essential property of all growing cells is that they must duplicate their DNA content exactly and ensure that each daughter cell receives an identical complement of DNA. This chapter deals with the mechanisms that ensure this equitable distribution of DNA to daughter cells.

8.2 The cell cycle and DNA synthesis

All growing cells undergo a cell cycle which is a term used to describe the time that elapses between successive cell divisions. It comprises two distinct periods:

- cell division and the separation of daughter cells;

- interphase, which represents a period of cell growth.

We can distinguish two levels of problems associated with ensuring that the daughter cells receive identical genomes (Note that we use the word genome to encompass all of the genetic information carried by a cell). One problem is, of course, the mechanism for replicating duplicating DNA. The other is the mechanism(s) that leads to both daughter cells each receiving a copy of the genome.

In this chapter, we have assumed that you already have knowledge of the structure of DNA and of the process of semi-conservative mode of DNA replication. We have therefore focused attention on the mechanisms underpinning the sharing out of the replicated DNA between daughter cells. If you do not know the basic structure of DNA or the process of semi-conservative replication, we recommend the BIOTOL text 'The Fabric of Cells'. We do however, briefly remind you, that DNA consists of two 'complementary strands' in which the nucleotide bases pair with each other in a

Margin notes: zygote; cell division, interphase; genome; semi-conservative; DNA replication; base pairing

particular manner. Thus adenine always pairs with thymine and guanine with cytosine. This arrangement provides the basis for the replication of DNA. Thus the two strands of DNA separate, each acting as a template for synthesis of a complementary strand. This is represented in Figure 8.1.

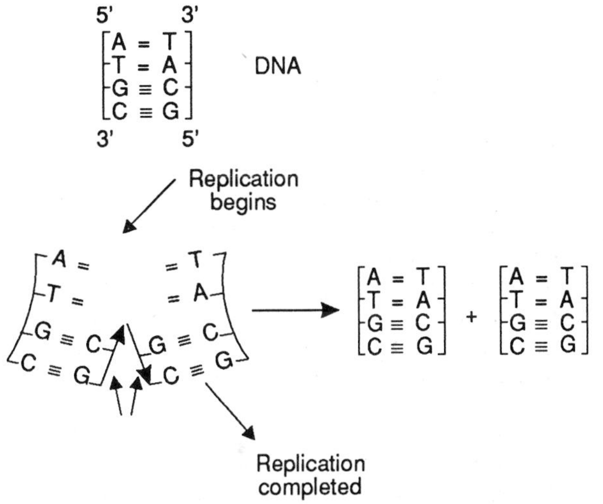

Figure 8.1 Semi-conservative replication of DNA.

In prokaryotic DNA, synthesis starts at a fixed point on the circular genome and takes place in the mesosome region.

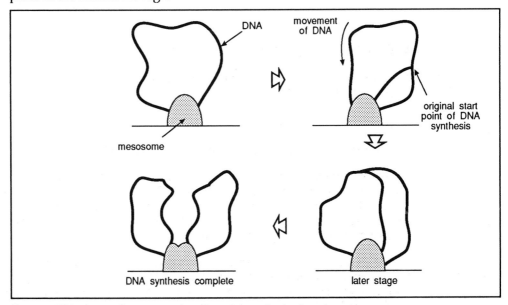

Figure 8.2 Diagrammatic representation of genome replication in prokaryotes.

We know that the synthesis of DNA requires the four deoxyribosyl nucleotide triphosphates (dATP, dGTP, dTTP, dCTP), DNA-dependent DNA polymerase and there is also an unwinding mechanism to open up short sections of the DNA helix so that the polymerase can copy the nucleotide strands (Figure 8.3).

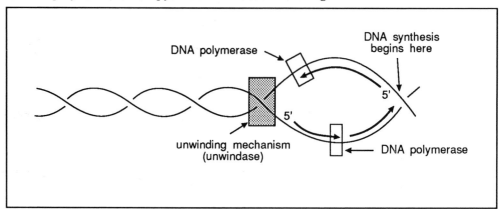

Figure 8.3 DNA replication fork.

replication fork

Note from Figure 8.3 that the DNA replication is asymmetrical. DNA strands are always synthesised in the 5' → 3' direction. Thus one strand can be made as a continuous strand while the other is made in short pieces which are subsequently joined (ligated) together.

In eukaryotes, the situation is rather similar but with some important differences. Remember that the DNA in eukaryotic genomes are wrapped around histones in looped domains. Clearly there must be some unwrapping to enable the DNA polymerases to gain access to the DNA.

multiple replication origins

Remember also that eukaryotes often carry very much more genetic information then do prokaryotes. A single replication fork in insufficient. Thus most eukaryotes have multiple origins for DNA synthesis (Figure 8.4).

Note that not all replication origins are activated at exactly the same time. We will see however that the replication of DNA is confined to a particular stage of the cell cycle.

8.3 The cell cycle in prokaryotes

Let us first deal with the cell cycle in prokaryotes because this is clearly the most straightforward. Prokaryotic cells contain a single, circular molecule of DNA. When bacteria are growing rapidly and dividing approximately every twenty minutes, DNA is replicated throughout most of the cell cycle. Soon after completion of DNA synthesis,

cytokinesis

the cell divides, a process termed cytokinesis. There does not appear to be an elaborate apparatus involved in cell division. The two daughter DNA molecules that result from DNA replication are attached at different points of the plasma membrane and this ensures that a copy of the DNA is delivered to each daughter cell upon cell division (see Figure 8.5). Certain bacteria have been shown to have special infolded regions of the

mesosomes

plasma membrane, termed mesosomes, which are believed to serve as anchor points for the attachment of the daughter DNA strands. We discussed mesosomes in Chapter 1.

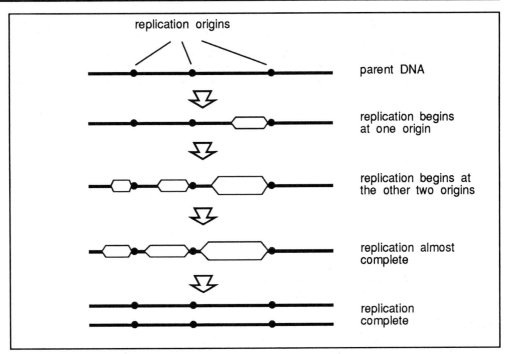

Figure 8.4 DNA replication in eukaryotes.

There is no condensation or de-condensation of DNA molecules during the prokaryotic cell cycle. Bear this in mind because it is in stark contrast to what happens in eukaryotic cells.

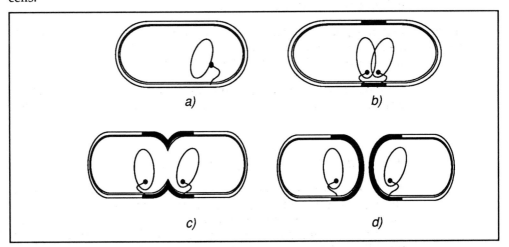

Figure 8.5 The prokaryotic cell cycle. DNA in the cell is attached to the cell membrane via the mesosome and remains attached during cell division. a) The circular chromosome which has already begun replication, is attached to the plasma membrane. b) When chromosomal replication is complete, the new chromosome has an independent point of attachment to the membrane.New membrane and cell wall form at the centre of the cell. c) More secretions of the cell wall form between the points of attachment of the two chromosomes.Part of this growth invaginates to give rise to a septum dividing the cell. d) Cell division is complete; each daughter cell has DNA attached to the membrane (see also Figure 1.26).

8.4 The cell cycle in eukaryotes

The division of a eukaryotic cell is a most spectacular phenomenon when viewed under the microscope. Two crucial events are clearly visible.

mitosis,
chromosomes

In the first event, mitosis, the contents of the nucleus condense to form visible chromosomes which are eventually collected into two equal sets.

cytokinesis

M-phase

In the second event, cytokinesis, the cell divides into two daughter cells in such a way that each daughter receives one of the two separated sets of chromosomes. Together these two events represent what is known as the M-phase (M = mitosis) of the cell cycle. In actual fact, only a small proportion of the cell cycle (approximately 10%) is taken up with the M-phase.

interphase

The other 90% of the time represents a phase which is known as the interphase. Microscopically, the interphase period appears to be a period of relative inactivity in which the cell simply grows in size. However, nothing could be further from the truth. More sophisticated techniques of observing the cell's activity show that, during interphase, the cell is making elaborate preparations for mitosis. Furthermore, these preparations occur in a well defined, ordered sequence.

8.4.1 Interphase

S-phase

In most cells, the replication of DNA is confined to a limited period within interphase known as the S-phase (S = synthetic phase). In this period each DNA molecule is replicated into two identical daughter DNA molecules.

gap periods

G_1

G_2

The S-phase is sandwiched between two gap periods known as G_1 and G_2. The S-phase is preceded by G_1 and followed by G_2. Other cellular components such as RNA, proteins and membranes are synthesised continually throughout the interphase. So, the order of events in the cell cycle is : M, G_1, S, G_2, M. There is no net synthesis of DNA in either G_1 or G_2 although damaged DNA can be repaired in either period. It follows from the above description that a cell contains two copies of its DNA molecules during G_2. A copy of each of these two DNA molecules passes into the daughter cells at the mitosis that follows G_2. The different stages of the eukaryotic cell cycle are illustrated in Figure 8.6. Look carefully at this figure. You will see a G_0-phase. Some cells leave the cell cycle and cease to divide. Such cells are said to be in G_0. Cells that enter G_0 generally acquire specific differentiated features. We will learn more of differentiation later.

8.4.2 Cell cycle times

Cells of different tissues, different species or different stages of embryonic development can have vastly different cell cycle times. Some of the fastest dividing eukaryotic cells are yeast cells and embryonic cells of organisms like the sea urchin and the frog. These cells can divide approximately every one to two hours. Most growing plant and animal cells take approximately ten to twenty hours to divide. Some cells take considerably longer. For example in cells of the adult human liver approximately one year elapses between successive M-phases. Some cells, like the terminally differentiated striated muscle and nerve cells never divide. In different growing mammalian cells the length of the M-, S- and G_2-phases is very roughly constant and the variation in the observed

quiescent

G_0-phase

cell cycle time is due solely to differences in the length of the G_1-phase. Non-dividing cells, like the striated muscle and nerve cells mentioned above, are generally arrested in a distinctive quiescent stage known as the G_0-phase.

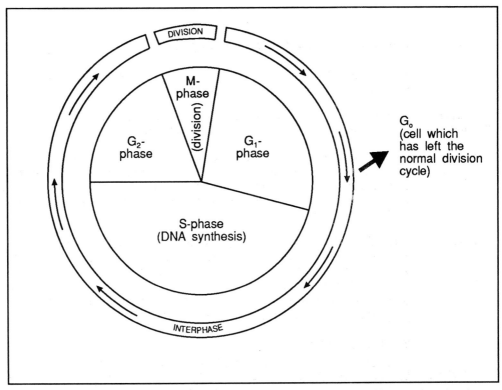

Figure 8.6 The four successive stages of a typical eukaryotic cell cycle. The M-phase consists of nuclear division (mitosis) and cytoplasmic division (cytokinesis). The two resultant daughter cells each enter the interphase of a new cell cycle. This begins with the G$_1$-phase during which the biosynthetic activity of the cell resumes following down regulation during M-phase. S-phase marks the onset of DNA replication during which each chromosome in the nucleus is copied. Cells leave the S-phase upon completion of DNA replication and enter G$_2$-phase which continues until mitosis starts, initiating the next M-phase. Cytokinesis terminates the M-phase and marks the beginning of the next cell cycle. The cell cycle times in eukaryotes vary widely from less than eight hours to more than a year. Most of this variation is the result of the variable lengths of G$_1$.

Now is the time to take a rest and to assess your progress by attempting to answer these two questions:

SAQ 8.1	Use the words below to fill in the blanks in the following statements. 1) The contents of the nucleus condense to form visible chromosomes at [ ]. 2) In the process of [ ] a cell splits into two daughter cells. 3) The easily visible events of mitosis and cytokinesis together occupy only a brief period of the cell cycle known as the [ ]. 4) The period that elapses between one mitosis and the next is known as [ ]. 5) The period of the cell cycle devoted to DNA synthesis is known as [ ]. Word Choice: M-phase, cytokinesis, interphase, mitosis, S-phase.

SAQ 8.2	Indicate whether the following statements are true or false. 1) During cell division the duplication of most of the cell's constituents need not be controlled exactly. 2) Cell cycle times vary from one cell type to another, with most of the variation occurring in G_1. 3) The M-phase of the cell cycle in rapidly growing cells can begin before the end of the S-phase.

8.4.3 The causal connections between the four phases of the cell cycle

It is now clear that the interphase is a period during which elaborate preparations are being made to allow the cell to divide and that these preparations occur in a precisely ordered sequence of events. The material in this next section covers some of the ways that the sequence of events can be investigated and discusses how the phases of the cell cycle are causally related.

cell culture

synchronised cells

It is difficult to analyse the events underlying the cell cycle in the complex tissues of an intact animal. Fortunately, certain cell types can be grown in isolation in a complex, semi-defined medium. The analysis of the events of the cell cycle has been simplified by the use of large populations of tissue culture cells which are in the same phase of the cell cycle. It is possible to obtain such synchronised cells in a number of ways. For example, the first attempts to prepare synchronous cultures involved the use of chemicals that were known to arrest cells at a particular stage of the cell cycle. Treatment of an asynchronous population of tissue culture cells will eventually cause all the cells to become arrested at the same stage. Upon subsequent removal of the chemical block all the cells will resume cycling at the same point in the cell cycle. You may recall from Chapter 4, that colchicine causes disaggregation of the microtubules of the cytoskeleton. These microtubules are also involved in the process of mitosis. Treatment of cells with this chemical therefore prevents cells carrying out mitosis. If we leave a population of cells in a medium containing colchicine, they will, therefore progress through the cell cycle until they reach mitosis. The cells thus all accumulate at the beginning of the

M-phase. If colchicine is then removed, in principle, the cells will all enter the M-phase together as the microtubules are re-established.

⊓ Examine Figure 8.6 again. Can you think of any other stage of the cell cycle that could be a target for producing a synchronous culture using an inhibitor?

Perhaps a good target would be DNA synthesis. An inhibitor of DNA synthesis would prevent cells from entering the S-phase of the cell cycle. Thus cell would accumulate at the beginning of the S-phase.

⊓ Are there any major problems in using inhibitors for producing synchronous cultures?

The simple answer is yes. Firstly the inhibitor has to be very specific for a single stage of the cell cycle. Secondly, when we remove the inhibitor, we are not truly examining a 'normal' cell cycle, but are, at least in part, examining recovery from the effects of the inhibitor. Despite these difficulties, however, some important and useful information can be gained from such studies.

Other methods of synchronisation involve more gentle procedures. Normally, tissue culture cells are grown in special plastic petri dishes to which the cells adhere tightly. However, during mitosis, there are changes in the cytoskeleton which cause the cells to round up and adhere weakly to the plastic culture dish making it possible to dislodge mitotic cells by gentle agitation. Mitotic cells collected in this way give a highly synchronous population which will immediately enter the G_1-phase (Figure 8.7).

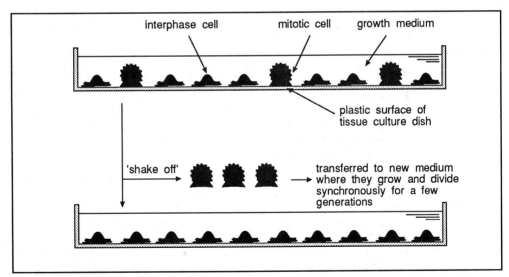

Figure 8.7. A gentle method for obtaining a synchronously growing population of animal cells in culture. Mitotic cells can be shaken off the dish on which they are growing. When transferred to a new dish, these cells continue to grow synchronously. There is however, some variability in the duration of each cell's cycle times, so, after a few divisions, synchrony is lost.

8.4.4 Critical cell cycle events take place against a background of continuous growth

continuous synthesis

What have synchronous cell cultures told us about the events of the cell cycle? One of the first investigations helped us map out the pattern of macromolecular synthesis during the cell cycle. For most of the cell's constituents growth is a steady continuous process which is interrupted only by the division of the cell. Thus the protein content increases more or less continuously throughout the cell cycle. Likewise, RNA synthesis proceeds at a steady rate, except during mitosis when the rate falls to near zero. This is presumably because the chromosomes undergo such condensation during mitosis that the DNA becomes inaccessible to the transcriptional machinery (Remember the description of DNA packaging in chromosomes dealt with in Chapter 6). We have already discussed one major exception to this picture of continuous cell growth and that is the restriction of DNA replication to the S-phase. A more detailed inspection of the synthesis of different types of protein also reveals that a relatively small number of proteins are not synthesised continuously throughout the cell cycle. Histones are the classic example of proteins that are only synthesised at a unique phase of the cell cycle. As we have learnt, DNA is complexed with histones to form chromatin. It is the condensation of the chromatin that will eventually give rise to the recognisable chromosomes. Newly replicated DNA needs a new complement of histones to form chromatin, it is therefore no surprise to find that histone synthesis is largely restricted to the S-phase of the cell cycle.

discontinuous synthesis

8.5 Control of the cell cycle

8.5.1 The restriction point in the cell cycle

A series of abrupt changes are now known to occur against this background of cell growth thereby triggering key events in the cell cycle. The onset of DNA synthesis at the beginning of the S-phase is the most easily detected. Others are not so readily identified. For most cells, there is a critical point in their G_1-phase at which the cell cycle becomes suspended if growth conditions are unfavourable. Once a cell passes beyond this point in the cycle, referred to as the restriction point or start, it undergoes some kind of internal change that commits it to complete the cell cycle according to the rigid sequence of events set out above.

restriction point

start

8.5.2 DNA synthesis is triggered by a cytoplasmic component - the S-phase activator

It is possible to fuse members of two different populations of cells by the addition of certain chemicals, for example polyethylene glycol, to the culture medium. The fusion of members of two populations of cells synchronised at different stages of the cell cycle has proved to be a very effective method of studying the cytoplasmic changes that trigger key events in the cycle. When an S-phase cell is fused with a cell in early G_1, the G_1 cell nucleus begins to replicate its DNA immediately. Thus the G_1 nucleus is apparently ready and able to replicate its DNA but has not yet generated the appropriate signal, or set of signals. The S-phase cell clearly has an abundance of the activating signal(s), as would be expected, and is able to activate the G_1 nucleus in the fused cell. The appearance of this S-phase activator, whose composition remains unknown, marks the boundary between G_1 and S.

S-phase activator

∏ Can you think of a way of determining whether the S-phase activator, responsible for initiating DNA synthesis, disappears after the cell completes DNA synthesis and proceeds into G_2?

This question can be easily answered by another type of cell fusion, this time the fusion of a cell in G_2 with a cell in G_1. The result is that the G_2 cell is unable to activate the G_1 nucleus to replicate its DNA, indicating that the S-phase activator, or at least some essential component, must disappear soon after cells enter G_2.

8.5.3 Mitosis is delayed until DNA replication is complete

M-phase
delaying factor

When a cell has passed through S and entered G_2 it will normally proceed through to mitosis at a fixed time. However, if DNA synthesis is blocked artificially, mitosis is delayed until after the block is removed and DNA replication is complete. Similarly, if an S-phase cell is fused with a G_2 cell, the G_2 nucleus is delayed in the cycle until the S-phase nucleus completes the replication of its DNA and catches up. Clearly, the cell must have an M-phase delaying factor, the level of which is dependent on whether the DNA is still undergoing replication. This signal could well be the persistent levels of the S-phase activator which, remember, does not decline until after DNA replication. The S-phase activator therefore may play a dual role. Increases in its concentration trigger the onset of DNA replication and the start of S-phase, decreases in its concentration could signal the transition from S to G_2.

8.5.4 Mitosis is triggered by an M-phase promoting factor

M-phase
promoting
factor

The disappearance of the M-phase delaying factor is not sufficient to induce mitosis. A further factor(s) is required. The existence of this so-called M-phase promoting factor (MPF) can be dramatically demonstrated by yet another cell fusion experiment. This time the two cells involved are an M-phase cell and a cell in any other phase of the cell cycle. When an M-phase cell is fused with a cell in any other part of interphase, the interphase nucleus is immediately induced to condense its chromosome in preparation for cell division. Of course, this type of fusion spells disaster for a cell that is either in G_1 or has yet to complete S because the premature condensation of its DNA will prevent further progress through the cell cycle. Under normal circumstances the MPF must be active only during a very small period in the cell cycle. The factors discussed above that delay the onset of mitosis (M-phase delaying factor) must be able to delay the appearance of the MPF. However, the dramatic result of fusing an M-phase cell with any other cell demonstrates that they cannot block its effects once it has been produced!

These experiments functionally define three diffusible control factors that govern the events of the cell cycle. The factors have to be diffusible, otherwise the cell fusion experiments would not work.

∏ Can you explain why the cell fusion experiments demonstrate that the control factors must be diffusible?

Basically, the answer is this. When we put a new nucleus into a different cytoplasm by a cell fusion technique, the nucleus responds to the condition of the new cytoplasm. Clearly, the signal must diffuse either directly or indirectly from the cytoplasm into the introduced nucleus.

The three diffusible control factors we have identified are:

- the S-phase activator, which is normally present only in S-phase cytoplasm and which is responsible for initiating DNA synthesis;

- the M-phase promoting factor, which is present only in the M-phase and which initiates chromosome condensation;

- the DNA synthesis-dependent M-phase delaying factor, which could possibly be identical to the S-phase activator.

| SAQ 8.3 |

Indicate whether the following statements are true or false.

1) In the G_1-phase cells undergo a critical transition called start, which is an internal change that marks the onset of DNA replication.

2) The rates of synthesis of many proteins are altered at specific stages of the cell cycle.

3) DNA replication is initiated by the S-phase activator.

4) DNA replication is terminated by the M-phase promoting factor.

∏ We would suggest you make a copy of Figure 8.6 and draw on when these three control factors are produced. It would help you to remember both the events of the cell cycle and the mechanism for controlling the cycle.

The causal relationship between these factors guarantees that the events of the cell cycle will always occur in a fixed sequence. DNA synthesis cannot be initiated until after the appearance of the S-phase activator. The S-phase activator and the M-phase delaying factor (which may be the same factor) cannot disappear until after DNA replication is completed and the M-phase promoting factor cannot be synthesised until after the disappearance of the M-phase delaying factor and the cell cannot pass through mitosis until the M-phase promoting factor has been produced. Clearly, the events of the cell cycle are linked together as a dependent sequence.

8.6 Mitosis

As the cell passes into the G_2-phase of its cell cycle it possesses two copies of each of its DNA molecules. The problem now facing the cell is to ensure that one complete copy of these DNA molecules is incorporated into each daughter cell following cytokinesis. This very precise distribution of the DNA molecules is achieved as a result of the process of mitosis. Following their replication in S-phase the two copies of each of the DNA molecules rapidly associate with the newly synthesised histones to form a DNA:protein complex known as chromatin which is dispersed throughout the nucleus. As the cell passes through G_2 the levels of the M-phase promoting factor increase and cause the dispersed chromatin fibres to begin to condense into thread-like chromosomes (remember the super-coiling described in Chapter 6). Each replicated DNA molecule forms a structure known as a chromatid upon condensation.

chromatin

chromosomes

chromatid

SAQ 8.4

Tick in the box provided the correct outcome of the [...] experiments.

1) The fusion of an S-phase cell with a G_1-phase cell

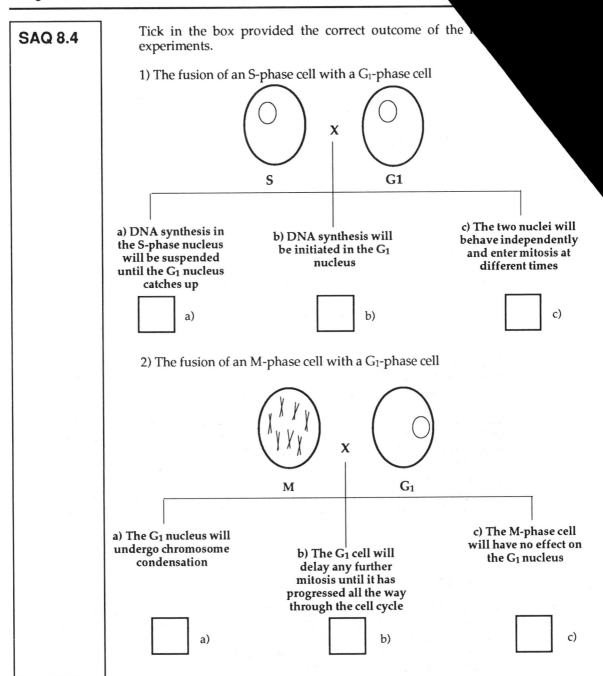

S   X   G1

a) DNA synthesis in the S-phase nucleus will be suspended until the G_1 nucleus catches up

☐ a)

b) DNA synthesis will be initiated in the G_1 nucleus

☐ b)

c) The two nuclei will behave independently and enter mitosis at different times

☐ c)

2) The fusion of an M-phase cell with a G_1-phase cell

M   X   G1

a) The G_1 nucleus will undergo chromosome condensation

☐ a)

b) The G_1 cell will delay any further mitosis until it has progressed all the way through the cell cycle

☐ b)

c) The M-phase cell will have no effect on the G_1 nucleus

☐ c)

sister chromatids

centromere

The pair of chromatids containing identical DNA molecules are known as sister chromatids. When sister chromatids are viewed under the scanning electron microscope they can be seen aligned and held together at a special region called the centromere which appears as a distinct constriction in electron micrographs. We have presented a diagrammatic version of sister chromatids hold together by centromeres in Figure 8.8.

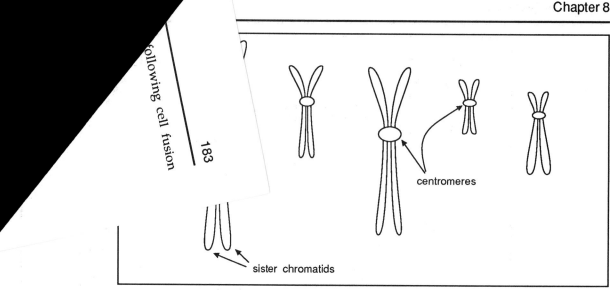

following cell fusion

183

Figure 8.8 A collection of chromosomes showing sister chromatids held together by centromeres. Note that the centromere need not be in the centre of the chromosome and that chromosomes are of variable length.

At the end of mitosis one of each of the sister chromatids will be allocated to each daughter cell. Remarkably, the chromatids themselves play no more than a passive role in this very precise redistribution. The active roles in this performance are played by two distinct cytoskeletal structures that transiently appear during the M-phase. The first player to appear is a bipolar mitotic spindle which is composed of microtubules and associated proteins. The mitotic spindle will be responsible for aligning the sister chromatid pairs along the equatorial plane of the cell and their eventual separation to opposite poles of the cell. This will ensure that each daughter cell receives a complete set of chromatids which are in fact the new cells' chromosomes.

mitotic spindle

The second player to take part in mitosis is a contractile ring which is made up of actin filaments and myosin. The contractile ring forms slightly later than the mitotic spindle just below the plasma membrane. Contraction of the ring pulls the membrane inward to divide the cell into two, thereby ensuring that each daughter cell not only receives one full complement of chromosomes but approximately half of the cytoplasmic components and organelles present in the parental cell. The formation of the mitotic spindle and the contractile ring is closely coordinated so that cytokinesis follows immediately after the end of mitosis. Because of their very rigid cell walls, plant cells require a different mechanism for cytokinesis and we shall discuss this later. We can summarise these events as in Figure 8.9. In this we have used only a single chromosome to represent the events.

contractile ring

Let us now examine the M-phase events a little more closely.

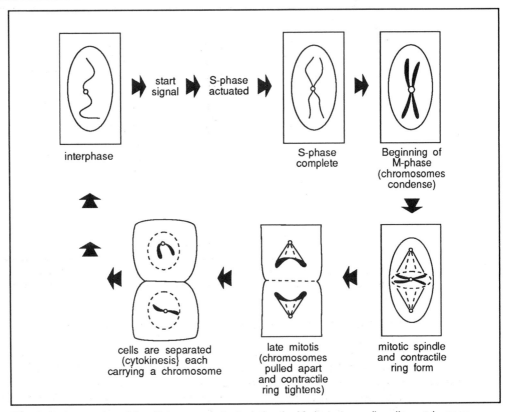

Figure 8.9 An overview of the chromosomal events during the M-phase (normally cells contain many chromosomes - Man has 46 chromosomes in each somatic nucleus).

8.6.1 M-phase (mitosis plus cytokinesis) consists of six stages

prophase

prometaphase

anaphase

telophase

The basic strategy of mitosis is remarkably constant among eukaryotic organisms. Of the six stages, the first five represent mitosis and the sixth represents cytokinesis. The five stages of mitosis are: prophase, prometaphase, metaphase, anaphase and telophase and they occur in this strict sequential order. Cytokinesis begins during anaphase and continues to the end of the mitotic cycle. The six stages of mitosis are shown schematically in Figure 8.10, together with a typical time course.

8.6.2 Formation of the mitotic spindle

Formation of the mitotic spindle in the M-phase of the cell cycle is accompanied by dramatic changes in the properties of cytoplasmic microtubules. So, before describing in detail the formation of the mitotic spindle, we provide a short revision of what you have learnt in Chapter 4 about microtubules.

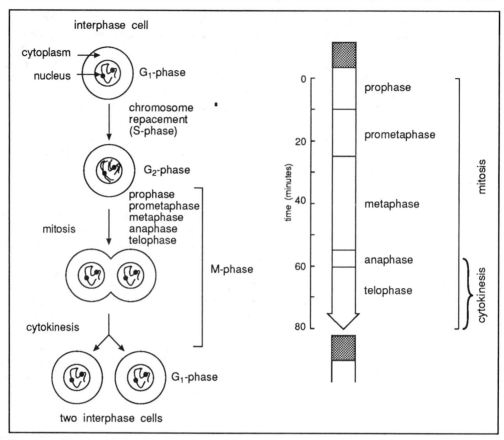

Figure 8.10 Duration of mitosis.

In Chapter 4 we described the cytoskeleton. Microtubules are just one element of the cytoskeleton. Cytoplasmic microtubules are hollow tubes made up by the association of tubulin heterodimers, dimers consisting of one molecule of α-tubulin and one molecule of β-tubulin. They can be seen to exist as single filaments that radiate from centres, so-called microtubule organising centres, lying close to the nucleus. The major microtubule organising centre in mammalian cells is the centrosome which is situated in the cytoplasm close to one side of the nucleus. The centrosome is responsible for organising the production of the mitotic spindle. It is as an area of rather amorphous material surrounding a pair of centrioles. (The situation is somewhat different in mitotic cells of higher plants where the microtubules end in an ill-defined region which is devoid of centrioles). Microtubules are rather labile entities but they can be stabilised by the binding of a special class of proteins, the microtubule-associated proteins. If you do not recall these features we suggest you re-read Chapter 4.

centrosome

centrioles

Let us examine how these features play a part in mitosis and cell division. The formation of the mitotic spindle begins way back in the S-phase when the centriole pair replicates whilst remaining associated with the centrosome. At the beginning of prophase the centrosome splits and each daughter centrosome becomes the focal point of a starlike aster of microtubules. The microtubules elongate and the two daughter centrosomes begin to move apart towards opposite poles of the cell. During the prometaphase the nuclear membrane breaks down allowing the mitotic spindle microtubules entry into the nucleus where they can make contact with the sister chromatids. Formation of the mitotic spindle is now nearly complete and the two daughter centrosomes constitute the two spindle poles. Use the description and figures provided to follow the events of mitosis.

Prophase

Microscopically, the transition from the G_2-phase to the M-phase is not a sharply defined event. In response to the M-phase promoting factor, chromatin slowly condenses into well-defined chromosomes, the number of which is characteristic of a particular species. Remember, each DNA molecule has duplicated during the S-phase and each copy of the resultant pair condense to give rise to sister chromatids. These are aligned along their length and joined at a region known as the centromere. The centromere is actually a specific DNA sequence which is required for the correct segregation of sister chromatids. The cytoplasmic microtubules begin to breakdown and are replaced by the formation of the mitotic spindle. The mitotic spindle is a series of microtubules that originate from the pair of centrosomes that resulted from centrosome duplication early in the S-phase. The daughter centrosomes begin to move apart, destined for opposite poles of the cell, and the microtubules that emanate from them continue to grow by the addition of tubulin.

Prometaphase

The onset of this phase is marked by the rapid dissolution of the nuclear membrane. This allows some of the spindle microtubules to enter the nuclear region. By now the daughter centrosomes have reached opposite poles of the cell. Specialised protein structures called kinetochores now begin to assemble at the centromeres of the sister chromatids. These kinetochores become attached to some of the spindle microtubules. At this particular point in time there are three distinct types of microtubules : 1) those attached to the kinetochore, known as the kinetochore microtubules, 2) the remaining microtubules in the spindle associated along the equator of the cell which are known as polar microtubules, and 3) the astral microtubules which lie outside the mitotic spindle. Each of the two sister chromatids becomes attached to opposite centrosomes via the kinetochore microtubules. The kinetochore microtubules extend in opposite directions from the sister chromatids in each chromosome and this exerts opposing tensions which cause them to move in an agitated fashion.

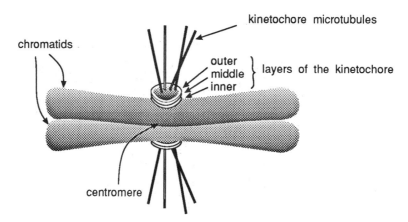

Metaphase

This agitated motion now causes the sister chromatids to become aligned in one plane along the equator of the cell, a region known as the metaphase plate. Each pair of sister chromatids are held in tension along the metaphase plate by the paired kinetochores and their associated kinetochore microtubules which are attached to opposite poles of the spindle.

Anaphase

Metaphase is a relatively stable state and under normal circumstances cells can remain in metaphase for an hour or more during which time the sister chromatids simply oscillate about the metaphase plate. Anaphase begins very abruptly when the sister chromatids separate synchronously. The trigger for sister chromatid separation is not precisely known. However, it is clear that the signal to initiate chromatid separation is not a force exerted by the spindle itself, since chromatids that are detached from the spindle begin to separate at the same time as those that are attached. There is some evidence that the trigger for chromatid separation involves a sudden increase in the concentration of cytoplasmic calcium ions. Once the sister chromatids have separated they move to opposite spindle poles where they will eventually be incorporated into the nucleus of the new cell. Their movement is the result of two independent forces within the mitotic spindle. First, the kinetochore microtubules begin to shorten. This is usually referred to as anaphase A. The second force represents the separation of the spindle poles as a result of the elongation of the polar microtubules and is known as anaphase B. Anaphase is short lived and lasts for only a few minutes.

Telophase

During telophase the separated chromatids reach the opposite spindle poles and the kinetochore microtubules disappear. The polar microtubules elongate further and a new nuclear membrane reforms around the two sets of daughter chromosomes at either pole. The chromosomes then begin to de-condense to become dispersed chromatin once more.

Prophase

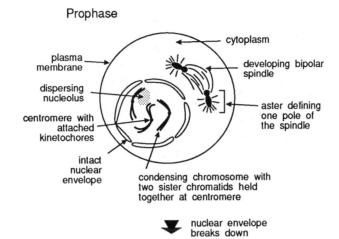

cytoplasm

plasma
membrane

developing bipolar
spindle

dispersing
nucleolus

aster defining
one pole of
the spindle

centromere with
attached
kinetochores

intact
nuclear
envelope

condensing chromosome with
two sister chromatids held
together at centromere

nuclear envelope
breaks down

Prometaphase

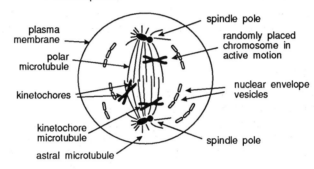

spindle pole

plasma
membrane

randomly placed
chromosome in
active motion

polar
microtubule

nuclear envelope
vesicles

kinetochores

kinetochore
microtubule

spindle pole

astral microtubule

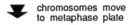
chromosomes move
to metaphase plate

Metaphase

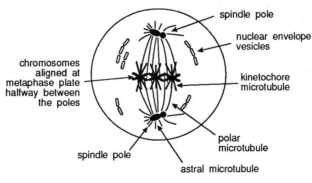

spindle pole

nuclear envelope
vesicles

chromosomes
aligned at
metaphase plate
halfway between
the poles

kinetochore
microtubule

polar
microtubule

spindle pole

astral microtubule

sudden separation of
sister kinetochores

Anaphase

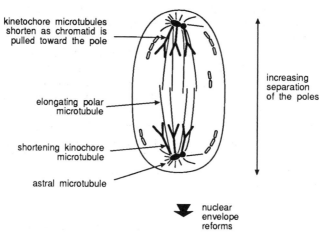

kinetochore microtubules
shorten as chromatid is
pulled toward the pole

elongating polar
microtubule

shortening kinochore
microtubule

astral microtubule

increasing
separation
of the poles

nuclear
envelope
reforms

Telophase

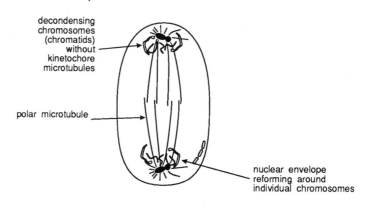

decondensing
chromosomes
(chromatids)
without
kinetochore
microtubules

polar microtubule

nuclear envelope
reforming around
individual chromosomes

cleavage furrow
splits cell in two

Cytokinesis

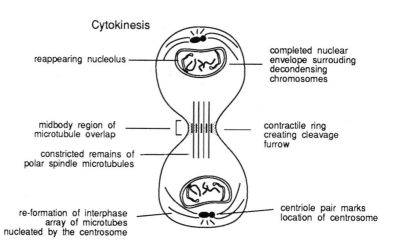

reappearing nucleolus

midbody region of
microtubule overlap

constricted remains of
polar spindle microtubules

re-formation of interphase
array of microtubes
nucleated by the centrosome

completed nuclear
envelope surrouding
decondensing
chromosomes

contractile ring
creating cleavage
furrow

centriole pair marks
location of centrosome

8.7 Cytokinesis

cleavage

During cytokinesis the cytoplasm divides by a process called cleavage. The mitotic spindle plays an important role in determining when and where this cleavage will occur. Cytokinesis usually begins in anaphase and continues through telophase. The cytoplasmic cleavage invariably occurs along the plane of the metaphase plate and at right angles to the long axis of the mitotic spindle. Cleavage is accomplished by the contraction of the contractile ring which is composed of actin filaments and myosin. There is compelling evidence to suggest that the force required is generated by the muscle-like sliding of the actin and myosin filaments in the contractile ring. The contractile ring does not get thicker as the cleavage furrow invaginates, suggesting that it must reduce its volume by shedding filaments. The ring is finally dispensed with when cleavage ends, as the plasma membrane of the cleavage furrow narrows to form the midbody, which contains the remnants of the polar microtubules packed together in a dense matrix.

midbody

cell plate

Because most higher plant cells are bounded by a rigid cell wall, cytokinesis in these cells is very different from the mechanism that has just been described. Cytokinesis in cells of higher plants is achieved by the construction of a new cell wall. The new cell wall, known as the cell plate, starts to assemble in a plane between the two new daughter nuclei. The position of this cell plate is associated with the residual polar microtubules which form a cylindrical structure known as the phragmoplast.

phragmoplast

| **SAQ 8.5** | Identify these four stages of mitosis: |

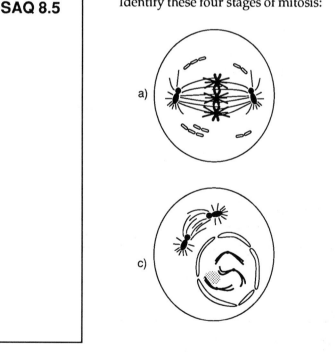
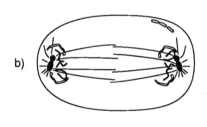
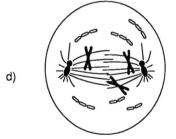

SAQ 8.6

Draw circles around the structures listed below that are involved in sister chromatid separation during mitosis.

Astral microtubules; kinetochore; polar microtubules; nuclear membrane; contractile ring; kinetochore microtubules.

8.8 Meiosis

Up until now we have not fully discussed the precise number of chromosomes present in a cell, except to say that the number of chromosomes per cell is characteristic of the species. So for example, Man has 46 chromosomes, dogs have 78 and corn has 20. In higher plant and animal species each cell contains two closely similar copies of each chromosome, termed homologous chromosomes, one member of the pair being inherited from the mother and the other from the father. In Man, therefore, the 46 chromosomes actually represent two sets of 23 different chromosomes. Cells that have two complete sets of chromosomes are said to be diploid. The vast majority of cells in higher plants and animals possess the diploid number of chromosomes. Cells of the human body therefore possess the diploid set of 46 chromosomes.

homologous
chromosomes

diploid

∏ Do all the cells of Man contain 46 chromosomes?

gametes

haploid

meiosis

bivalent

There are notable exceptions to this rule especially represented by the reproductive cells which are called gametes (the sperm and the egg). Gametes only possess a single copy of each chromosome. In humans this means that gametes possess only 23 chromosomes, the so-called haploid number of chromosomes. Yet the cells that give rise to the haploid gametes are themselves diploid. Clearly their division to produce gametes cannot follow the usual route of mitosis described above, because, as we have seen, mitosis involves the faithful replication of the DNA content and cannot therefore lead to a reduction in the number of chromosomes. The cell division that leads to the production of haploid gametes from diploid cells is highly specialised and is termed meiosis.

Meiosis involves two separate cell divisions and yields four haploid cells from a single diploid precursor. Before meiosis begins, the DNA in the homologous chromosomes is replicated, just as in mitosis, to yield sister chromatids. These sister chromatids remain attached at their centromere. The net result of this replication of homologous chromosomes is the production of two sets of sister chromatids, representing four closely similar copies of the same chromosome. The two sets of paired sister chromatids derived from the two homologous chromosomes behave as a single unit, called a bivalent, during the first of the two meiotic divisions. Just as in mitosis, a spindle is produced which causes the bivalents to line up along the equatorial plane. The sister chromatids remain firmly attached at their centromeres. The sister chromatid pairs, still firmly attached, then move to opposite spindle poles and the cell divides to give two daughter cells each of which possesses a set of sister chromatids, still firmly attached, derived from one of the two homologous chromosomes.

This first meiotic division is followed by an interphase period which unlike the interphase in mitosis, does not involve any DNA replication. During the second meiotic division that follows the sister chromatids align on a second spindle and separate, as in normal mitosis, to opposite spindle poles. Cytokinesis then follows to produce cells that now possess the haploid number of chromosomes. Meiosis thus consists of two nuclear divisions following a single phase of DNA replication so that four haploid cells are produced from each cell that enters meiosis.

The process of meiosis is outlined below.

1) interphase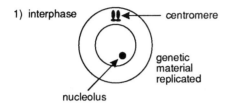

1) During the first interphase period the genetic material is replicated. Each chromosome of the homologous pair is replicated.

2) early prophase I

2) Early prophase I. Chromosome replication is complete. The sister chromatids remain together as a pair, but will not become readily distinguishable until late prophase I.

3) middle prophase I

3) Middle prophase I. The sister chromatids of homologous chromosome pairs align along their entire lengths and become attached to the spindle. They also condense.

4) late prophase I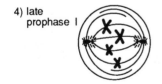

4) Late prophase I. Sister chromatids of each of the homologous chromosomes become visible. These four chromatids remain together and function as a unit, the bivalent. Bivalents begin to move toward the cell's equator.

5) metaphase I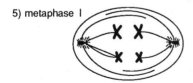

5) Metaphase I. The bivalents align themselves along the equator.

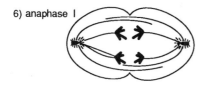

6) anaphase I

6) Anaphase. The sister chromatids remain attached at their centromeres, but the bivalents separate and the sister chromatids derived from one homologous chromosome move to one spindle pole, whilst the sister chromatids derived from the second of the homologous chromosomes moves to the opposite spindle pole.

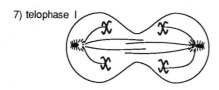

7) telophase I

7) Telophase I. A nuclear membrane forms around the complement of sister chromatids at each of the pole. Each sister chromatid pair remains firmly attached. Cytokinesis then follows.

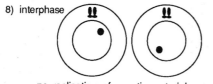

8) interphase

no replication of genetic material occurs

8) Second interphase. The sister chromatid pairs begin to de-condense. There is no further chromosome replication.

Second meiotic division

9) prophase II

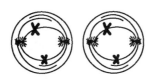

10) metaphase II

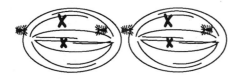

11) anaphase II

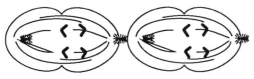

12) telophase II

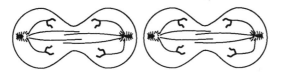

Steps 9) to 12). These events are similar to mitosis. In prophase II 9) the sister chromatids condense and become attached to the second meiotic spindle. These align along the equator of the cell during metaphase II 10). The sister chromatids finally separate and move towards opposite spindle poles in anaphase II 11). At telophase II 12) one of the sister chromatids is delivered to each daughter cell. The result of meiosis is four haploid cells derived from a single diploid precursor.

SAQ 8.7

The haploid chromosome number of the house mouse, *Mus musculus*, is 20. Remember, 20 chromosomes represents 20 molecules of DNA. Insert in the space provided the number of DNA molecules in each of the following mouse cells.

1) Adult liver in the G_1-phase.

2) A sperm cell.

3) A cell in prophase I of meiosis.

4) In each of the daughter cells following the first meiotic division.

5) A fertilised mouse egg.

8.9 Organelle production

This chapter has been predominantly concerned with examining the mechanisms that ensure that both daughter cells receive appropriate copies of the genetic material. It is, however, important that the daughter cells also receive copies of the major organelles. In this section, we will focus on the production of mitochondria and chloroplasts.

organelle

DNA

Both these organelles contain DNA which code for a limited number of proteins. They also contain ribosomes and are capable of protein synthesis. We can therefore visualise two sets of proteins in these organelles, namely:

- those which are coded for by the nuclear DNA and those which are produced in the cell cytoplasm;

- those which are coded for by the organelle DNA and which are produced in the organelle.

The organisation of the DNA, the nature of RNA synthesis and the structure of the ribosomes in mitochondria and chloroplasts are typical of those found in prokaryotes. Thus the DNA is a single circular molecule (although there may be more than one copy per organelle) and the DNA-dependent RNA polymerase and ribosomes show the same patterns of inhibition by antibiotics as those found in prokaryotes.

∏ Can you think of why these molecular biological features of mitochondria and chloroplasts should be similar to those of prokaryotes?

The answer is that these organelles have evolved from prokaryotic cells which have become endosymbionts in eukaryotic cells (If you do not remember this point, we suggest you refresh your memory by reading the final sections of Chapters 2 and 5).

organelle

free cells

New chloroplasts and mitochondria never arise *de novo*. They are produced by growth and division of existing organelles. The simple image is to think of them as prokaryotic cells (without cell walls) growing and dividing within a host cytoplasm. There is however, a very critical point to realise: how is organelle growth and division regulated such that not too many or too few are produced? If too few were produced, then gradually the host cell line, as it grew and divided, would become depleted of the organelle. Experimentally, it is possible to restrict organelle development and to produce cells which do not contain either chloroplasts or mitochondria.

∏ What sort of strategy could be adopted to produce cells without mitochondria or chloroplasts?

The usual approach is to use an inhibitor which blocks organelle protein synthesis without inhibiting nuclear/cytoplasm mediated protein synthesis. Typical of these inhibitors are the antibiotics that can be used to inhibit bacterial protein synthesis (eg chloramphenicol, erythromycin, tetracycline). In carrying out such experiments, it must be remembered that the host cells must be provided with suitable nutrients which will enable them to grow without mitochondria or chloroplasts.

organelle DNA
synthesis

Exactly how organelle growth and division is regulated, is not fully understood. We know, for example, that each organelle usually contains several copies of DNA (5-50 per organelle) and that there may be many organelles per cell ($1-10^7$ per cell). We also know that organelle DNA synthesis is not restricted to the S-phase (ie nuclear DNA synthesis phase). It appears that individual organelle DNA molecules are replicated more or less at random. Some being copied many times, some not at all, during each cell cycle.

Despite this apparent randomness, the total number of organelle DNA molecules and organelles doubles every cell cycle. Daughter cells receiving copies of the organelles at each division.

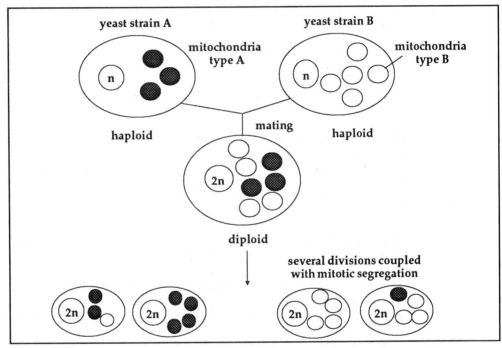

Figure 8.11 Mitotic segregation in yeast. Note that the different types of mitochondria depicted by ● and ○ are not shared equally in the progeny cells.

mitotic
segregation

Interestingly, there is substantial evidence that in many systems (eg yeasts), the mitochondria are not shared randomly during mitosis. The evidence for this is shown diagrammatically in Figure 8.11. The process is called mitotic segregation.

In mammals, all the mitochondria are inherited from the female (Figure 8.12).

Thus, the picture that has emerged is that chloroplasts and mitochondria can grow and divide in a manner not unlike that observed in prokaryotes, but exactly how this growth and division is regulated or how the progeny organelles are segregated at the time of cell division still remain unsolved mysteries.

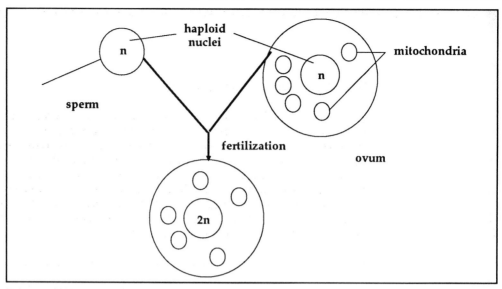

Figure 8.12 Maternal mitochondrial inheritance in mammals.

Summary and objectives

This chapter has dealt with the growth and division of cells. We learnt that in prokaryotes the process is relatively straightforward compared to that in eukaryotes. We have learnt that the cell cycle in eukaryotes can be divided into well organized phases and that these phases are carefully regulated by three types of cell cycle regulators. We have also learnt that two types of division take place in eukaryotes, namely mitosis and meiosis. The latter division is a highly specialized process in which diploid cells divide to form haploid gametes. We have also examined the production and partitioning of the major cell organelles.

Now that you have completed this chapter you should be able to:

- describe the stages of the eukaryotic cell cycle, emphasizing the strict sequential order in which these stages occur;

- explain how intracellular control mechanisms ensure that this sequential order is maintained;

- describe in outline, the kinds of experiments which led to the discovery of cell cycle regulators;

- describe the events that occur during mitosis and recognize cells that are at particular stages of mitosis;

- explain the unique nature of meiotic cell division;

- describe the events that take place during meiosis and recognize cells that are at particular stages;

- indicate the major differences between mitotic and meiotic cell division;

- describe how chloroplasts and mitochondria are produced and explain the unsolved issues concerned with organelle growth, division and segregation.

From single cells to multicellular organisms

From single cells to multicellular organisms

9.1 Introduction

Unicellular organisms like bacteria have been extremely successful in their ability to adapt to different environments. The fact that over half of the Earth's biomass is represented by unicellular organisms attests to their great versatility to colonise the many varied environments represented on earth. By definition, unicellular organisms are totally self-sufficient in the sense that they can either obtain from their environment, or synthesise, all of the substances required for growth and replication.

If unicellular organisms are so successful at survival, which they clearly are, why do you think that there has been a selective advantage for the evolution of more complicated multicellular organisms? The answer almost certainly relates to the fact that the evolution of multicellularity permits cells within the organism to acquire specialised functions. Furthermore, specialised cells within the organism can co-operate with one another. This specialisation and co-operation enabled multicellular organism to exploit resources that no single cell could utilise so well.

specialisation

co-operation

9.2 The association of unicellular organisms into colonial forms

An early step in the evolution of multicellularity was almost certainly the association of unicellular organisms into multicellular colonies. Indeed this process of colonialism is not unique to the eukaryotic world. Certain prokaryotic cells display rudimentary social behaviour. The Myxobacteria, for example, stay together in loose colonies where individual cells of the colony synthesise and secrete digestive enzymes into a common pool. This increases the efficiency of feeding on the insoluble organic molecules present in the soil in which the colony lives. A clear example of their ability to adapt to changes in their environment occurs when their food supply runs out. This triggers a tight association between the bacteria of the colony by a process of cell aggregation to form a structure known as a fruiting body (Figure 9.1). Once the fruiting body has formed, the bacteria begin to differentiate into spores that are designed to withstand the deprivations that initiated this process. By differentiation, we mean that the cells take up a different form. Thus although they contain the same genetic information, they become phenotypically different. Once the conditions become more favourable, the spores can germinate to produce a new colony of bacteria.

colonialism

aggregation

fruiting body
spores

differentiation

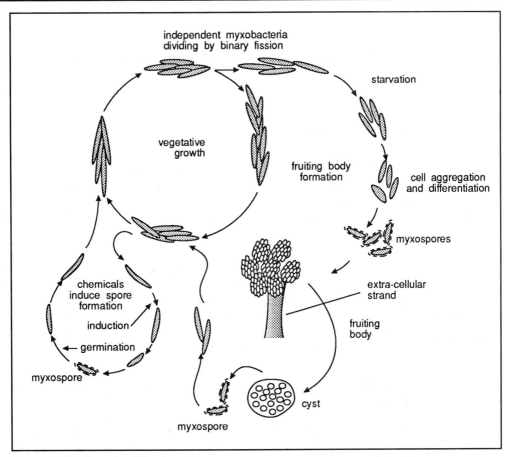

Figure 9.1 The life cycle of Myxobacteria.

9.2.1 Green algae

Some of the earlier eukaryotic cells to evolve, for example the green algae, exist in unicellular, colonial and multicellular forms, depending on precisely which type we are considering. The increasing complexity of this organisation possibly reflects the progression that occurred during the evolution of higher plants and animals. *Chalmydomonas*, for example, is a unicellular green algae which possesses both chloroplasts and flagella (Figure 9.2). In closely related genera, groups of these flagellated cells exist in colonies which are held together by molecules which are elaborated and secreted by the cells of the colony. These colonies can vary from a relatively small number of individual cells (4,8,16 and 32), as in the genus *Gonium*, to as many as 50,000 or more, as in *Volvox*. In the smaller colonies the flagella of individuals beat independently, but since all the flagella are oriented in the same direction the net result is that the colony can be propelled through the water. In *Volvox* the cells become linked together by extremely fine cytoplasmic bridges to form a hollow sphere. These connections allow the beating of the flagella of individuals to be coordinated. In other words, the cells in this particular type of colony can actually communicate with each other. There is also some *division of labour* within the *Volvox* colony. Certain cells which are collected together at one end become specialised for reproduction and serve as precursors of new colonies. So dependent are individual members of the colony on the others that none can survive independently and disruption of a colony results in the death of all the individual cells.

division of
labour

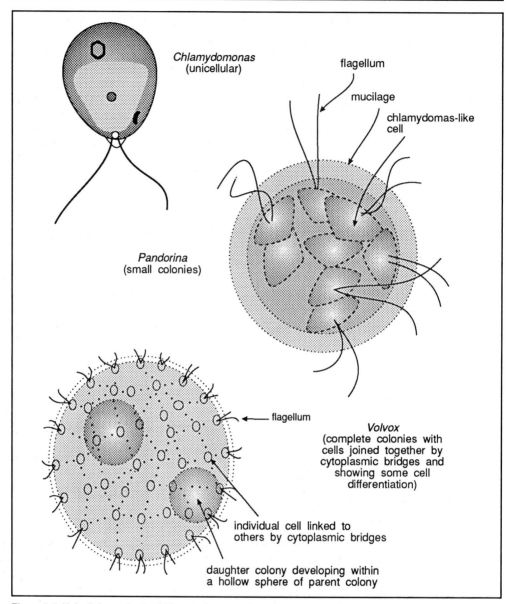

Figure 9.2 Unicellular and colonial forms of green algae (stylised).

Volvox is more like a multicellular organism than just being simply a colony of identical cells. For example, all of its flagella beat synchronously, the colony consists of structurally and functionally distinct cells and it can swim toward a distant source of light. As we mentioned above, the reproductive cells are located toward one end of the colony where they actually divide to form microcolonies which are harboured within the parental colony before being released into the environment. Thus the cells of *Volvox* become specialised and co-operate, both essential features of a multicellular organism. Perhaps in *Volvox* we are beginning to see evolution's basic blueprint for a multicellular organism.

9.2.2 The role of cell cohesion in the development of multicellularity

Multicellularity depends on the ability of the individual component cells to stick together. Several strategies have developed throughout evolution to bring about this cell cohesion. In *Volvox* the individual cells that form the sophisticated colony do not completely separate following cell division, instead daughter cells remain attached via fine cytoplasmic bridges. In higher plants, the cells not only remain in contact via cytoplasmic bridges, called plasmodesmata (see Chapter 2), but they are also encased in a rigid cellulose cell wall. Cells of animals do not possess cell walls and it is rare to see cytoplasmic bridges connecting cells. Instead, the cells are held together by a mesh-like structure called the extracellular matrix. We will discuss the composition and the role of the extracellular matrix in later sections.

plasmodesmata

ellulose cell
wall

extracellular
matrix

9.2.3 Epithelial sheets

cell
re-arrangement

If a sponge is disaggregated into individual cells by passage through a very fine sieve, the cells will commonly reaggregate into an intact sponge. Initially the disaggregated cells reform a large mass which rearranges into a sheet of cells called epithelia. This epithelial arrangement of cells is perhaps the single most common way that animal cells are assembled into multicellular structures. We can show you the importance of epithelia by discussing yet another lowly group of the animal kingdom, the coelenterates, which includes sea anemones, jelly fish and the freshwater *Hydra*. Members of this particular group are the first of the animal kingdom to possess a nervous system, although it is very primitive compared to that present in the higher eukaryotes, like mammals. The structure of the coelenterates can best be discussed with reference to *Hydra*.

9.2.4 *Hydra*

ectoderm

endoderm

coelenteron

Hydra are constructed from two layers of epithelia (Figure 9.3). The outer layer is called the ectoderm and the inner layer is called the endoderm. The endoderm surrounds a cavity known as the coelenteron which serves as an area in which food is digested. The food enters the coelenteron through a mouth which is situated at the head of the *Hydra*. These two epithelial sheets not only give *Hydra* mechanical support but they also give it a very primitive digestive tract. The endoderm is an effective barrier to the diffusion of molecules which ensures that the coelenteron provides a closed chemical environment suitable for the *Hydra's* digestion.

The endodermal cells become specialised to perform different functions. Some synthesise and secrete digestive enzymes into the coelenteron where they help digest the food that enters via the mouth. Other cells actually absorb the partially digested products of the coelenteron enzymes and digest them further. The ectodermal cells also become specialised for their task of interacting with the outside world. For example, the ectoderm contains specialised cells that can paralyse the small animals that serve as food for the *Hydra*. In between these epithelial layers lies a very narrow layer which houses the nerve cells that run the entire length of the body. The nerve cells are in intimate contact with the cells of the endoderm and the ectoderm whose activities they control. A detail of a portion of the body wall of *Hydra* is given in Figure 9.4. We would not expect you to learn all the names of the various types of cells but we would like you to recognise that even in these primitive animals, there has been an enormous amount of cell specialisation. Thus the musculoepithelial cells of the ectoderm enable the *Hydra* to expand or contract itself; the nemoblasts have the sting for stunning prey (each *Hydra* might have several different types of nematoblasts) and the nerve cells allow 'communication' between cells at one end of the animal to the other.

nemoblasts

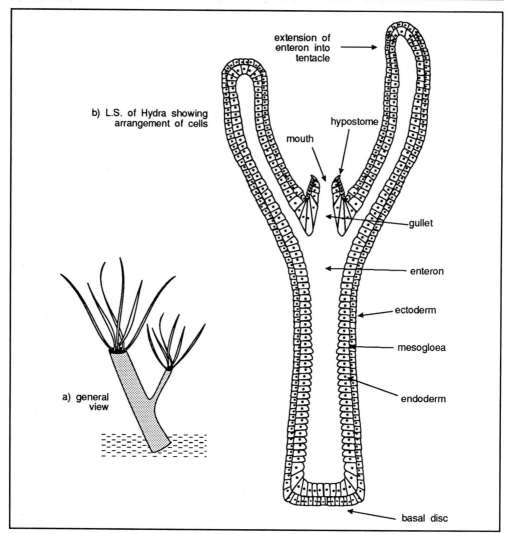

Figure 9.3 Diagrams of the body of *Hydra.* NB mesogloea is the layer between the ectoderm and the endoderm.

If we added to this complexity, the knowledge that *Hydra* may reproduce sexually, we immediately recognise that there must be even further cell differentiation to produce the gametes (ova and sperm) which will probably be produced in special structures. *Hydra* are also capable of 'walking', looping along, rather like a caterpillar (Figure 9.5).

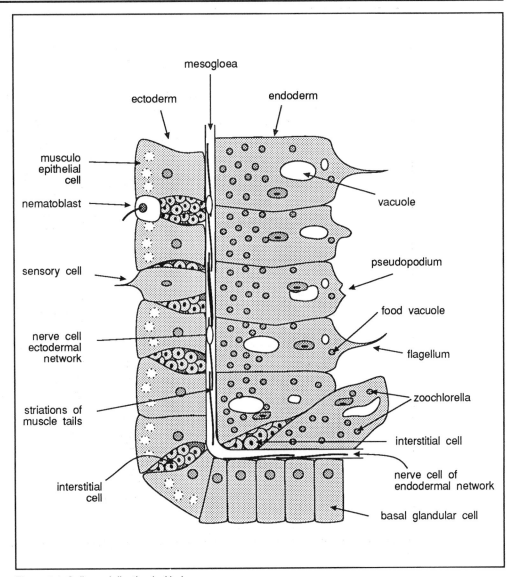

Figure 9.4 Cell specialisation in *Hydra*.

∏ From this description of *Hydra*, what are the special features of multicellular animals? (Jot down a list before reading on).

You could list many features, but the main ones are: many different types of cells are produced; each cell type is produced in the correct functional position and in the correct proportion; cells act co-operatively. Thus prey-stunning cells are produced on the outside (ectoderm), food ingesting cells on the inside and, for movement to occur, the individual cells have to act together.

The *Hydra* has served us well to illustrate these points, but is what we have learnt about *Hydra* relevant to higher organisms? First, let us think about the co-operation between cells, this means that cells need to communicate with each other.

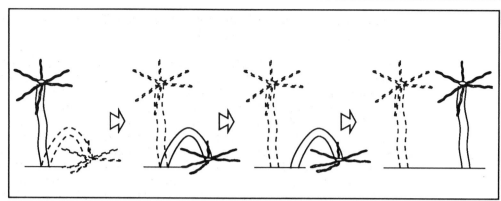

Figure 9.5 *Hydra* movement.

9.3 Cell-to-cell communication

We have learnt that the cells of *Hydra* can communicate with one another. This can also be demonstrated by surgically removing one end of the *hydra*. The remaining cells react to this by rearranging themselves, in some cases changing their specialist function, in order to reform a complete, although smaller, animal. Signals must pass from cell to cell in order that this rather precise reaction to amputation can occur.

The vastly more complex higher animals have evolved from the rather humble coelenterates. However, the principles of cell co-operation have been exploited in the evolution of these higher animal forms. Epithelial sheets of cells line all cavities within the body to create precise environments in which the differentiated cells can best perform their functions. Specialised cells also interact and communicate with one another via chemical signals which serve to control and co-ordinate the functions of the various parts of the animal. However, as higher and higher animal forms have evolved so too have the numbers of specialised cell types that an individual possesses.

Now is probably an appropriate time to discuss the generation of these specialised cell types by a process known as cell differentiation.

9.3.1 The development of complex patterns of differentiated cell types

Although in the course of evolution multicellular organisms have taken on more and more complex forms, each organism usually develops from a single cell, the fertilised ovum. The many cells of the mature organism are derived by repeated division of this single precursor cell. The vertebrate body contains more than 200 distinct cell types, each specialised to perform a particular function. Yet all of them are derived from the fertilised ovum. As this single cell grows and divides the cell number obviously increases. However, running in parallel to this increase in cell number there is a process going on, known as cell differentiation, which results in groups of cells adopting particular structures and functions.

The differentiation of cells is not the result of changes in the DNA content of the cells. Nearly all differentiated cells have exactly the same DNA content. It is produced by the differential usage of the information that is embedded in the DNA. In other words, cell differentiation depends on the activation and repression of sets of genes (ie there is differential gene expression). Different pathways of cell differentiation result from the

ovum

differential
gene
expression

activation and repression of different sets of genes. The initial cues for these altered patterns come from neighbouring cells.

However, you have to realise that cell differentiation is occurring against a background of cell division and growth in the number of cells. If cell differentiation and development is to be successful, it is essential that once a cell has become programmed to produce a particular cell type, it can pass this programme on to its daughter cells upon cell division. Thus the development of complex patterns of differentiated cell types not only depends on the ability of cells to be programmed at the level of DNA, but it also depends on the cell having a memory of that initial development response so that it can be passed on to its daughters at cell division. A consequence of this process is that a cell's final character is not only fashioned by the final environment in which it finds itself, but it is also dependent on the sequence of influences to which it has been exposed during the course of development.

cell determination

If a cell within a developing embryo goes through a stage in which its destiny is fixed it is said to be determined. The original cue for this determination would almost certainly have been environmental and resulted in alterations in the patterns of genes that are used by the determined cell and its progeny. The determination of a cell need not always lead to a detectable change in the appearance of the cell. However, determination initiates a programme of events that will eventually lead to the cells derived from the first determined cell acquiring a particular characteristic or set of characteristics. The final expression of this determined state, which again is a consequence of yet other environmental signals, is termed cell differentiation.

cell differentiation

totipotent cells

If you think of a fertilised ovum, you will realise that all of the cells of the organisms will develop from this ovum. We can say therefore, that an ovum is totipotent. We have seen that later in embryo development, some of the daughter cells produced, loose this totipotency and become 'determined' (committed to become a particular cell type or a small selection of cell types). We can describe a cell which has lost its totipotency but can still produce a variety of cell types as pluripotent.

pluripotent cells

SAQ 9.1

From the statements below, which of the cell types can be said to be displaying:

1) totipotency, 2) pluripotency, 3) commitment to being one cell type.

a) Plant cambial cells, isolated and grown in tissue culture can be stimulated to produce new plantlets by using appropriate incubation conditions.

b) Red blood cells ultimately die without undergoing further cell division.

c) Cells of the 'colony forming units' of bone marrow give rise to leucocytes and red blood cells but cannot produce new epithelial cells.

One of the striking features of animal development is that animals which are clearly different in their mature forms are often very similar during their early developmental stages. In other words, basic developmental programmes tend to be conserved in evolution.

Π Why do you think that basic developmental programmes have been conserved in evolution?

This is not an easy question to answer. What you need to think about is how evolving animals actually acquire new characteristics. They do this by altering the genetic messages that are embedded in the base sequence of a particular gene. This in turn allows the synthesis of a novel protein which will generate a new, or improved, characteristic. Evolution of animals is driven by alterations in genes which result in new cellular characteristics that generate selective advantages in organisms in which they have occurred. The cells of the early developing organism and their progeny are to be the precursors for many different differentiated cell types in the mature animal. It is likely that alterations in the genes of these early cells will almost certainly affect the activity of many different mature cell types and, whereas the mutation may be beneficial for some of these, it may also be detrimental to others. Thus it would appear that only the cells that are near to the end of a particular programme of differentiation have the freedom to acquire new gene functions via mutation since the only cells affected will be the cells of that particular pathway. The basic strategy of cell development thus appears to have been established relatively early during evolution and that acquisition of new characteristics that have allowed new species to evolve is the result of genetic changes that occur in cells that are very close to the end of a developmental programme.

9.4 The cells of the vertebrate body

tissue

organ

There are over 200 distinct cell types in vertebrates, including humans. These cells are organised in a highly complex but very specific manner into tissues and organs. A tissue is a collection of co-operating specialised cells which constitute a functional unit within an organism. An organ is a larger functional unit which consists of different types of tissue. A tissue virtually always consists of more than one type of differentiated cell.

∏ Why do you think that tissues usually consist of more than one cell type?

stability

nutrition

communication

immunity

The answer to this question is connected with the fact that, despite the great variety of tissues, a number of general demands can be formulated which apply to every tissue type. These demands include stability (this is provided by the extracellular matrix), nutrition (which requires the formation of capillary blood vessels by vascular epithelium), communication (achieved by the axons of the nerve cells), immunity (provided by macrophages and lymphocytes of the immune system) and pigmentation (caused by the presence of melanocytes). All of these cell types, which are often formed outside the tissue, have to penetrate and cross a tissue during its growth and development.

turnover of cells

homeostasis

With the exception of a few cell types, the differentiated cells of all tissues must be renewed continually. Old or damaged cells die, are removed and have to be replaced by new cells, a process known as cellular turnover. The exceptions to this general rule are cells of nervous tissue, heart muscle and the eye lens which, after their formation in the early embryonic stages, are incapable of cell division or replacement with new differentiated cells. In an adult, the rate of renewal and breakdown of cell types is in perfect balance. This is called homeostasis.

duplication
stem cells

The supply of new differentiated cells to replace those lost through natural wear and tear can be achieved in two ways:

- they can form by simple duplication of existing differentiated cells to give two identical daughter cells possessing the same differentiated state as the parental cell;

- they can be generated from undifferentiated stem cells, a process which requires that the products of stem cell division undergo further differentiation to produce the character of the cells being replaced.

SAQ 9.2

Why should the rate of cellular turnover of intestinal epithelium be higher than that of pancreatic cells?

9.4.1 Renewal by simple duplication

We will explain the strategy for tissue renewal by cell duplication using the liver as an example. Virtually every nutrient taken into the blood from the intestine first passes through the liver before being forwarded to other tissues and organs. Clearly, being at the forefront, as it were, the liver has a high risk of damage due to toxic substances in the blood. Indeed, detoxification of otherwise toxic compounds is an important defensive role played by the liver. There is therefore a need for continual replenishment

liver cells

of liver cells that become damaged and die. This, nevertheless, represents a fairly slow rate of cell turnover. However, liver, like many tissues whose normal rates of cell renewal are slow, can be stimulated to produce new cells at higher rates when the need arises. For example, if two-thirds of an adult liver is surgically removed, a liver of nearly normal size will be generated from the remaining remnants within a week or so. This regeneration process is totally dependent on simple cell division of the remaining liver cells.

hepatocytes

sinusoids

endothelial cells

Kupffer cells

fibroblasts

Like most tissues, the tissue of the liver is a mixture of different cell types (Figure 9.6). The majority of the dietary functions of the liver are performed by the hepatocytes. These are arranged in folded sheets which face blood-filled spaces called sinusoids. The hepatocytes are separated from the blood by a thin layer of flattened endothelial cells. In addition, the liver contains both specialised macrophages, called Kupffer cells, which engulf particulate matter in the blood, and a small number of fibroblasts which provide a supporting framework of connective tissue.

cirrhosis

All four cell types are capable of regeneration by simple cell division. Indeed, all need to regenerate if an intact functional liver is to be formed following the surgical removal of the majority of the organ. However, for perfect regeneration their multiplication must be tightly controlled. An imbalance in the regeneration programme can often have very serious consequences. This may be illustrated by the case of constant alcohol abuse. The toxic nature of the alcohol kills the hepatocytes and persistent exposure of the liver to alcohol results in an inability of the hepatocytes to recover to their normal levels within the liver. The fibroblasts take advantage of this situation and grow to colonise the space that would normally be occupied by the hepatocytes. This results in the liver becoming irreversibly clogged with connective tissue leaving little physical space for the hepatocytes to grow, even after the withdrawal of the alcohol. This clogging is the cause of the cirrhosis that accompanies chronic alcoholism.

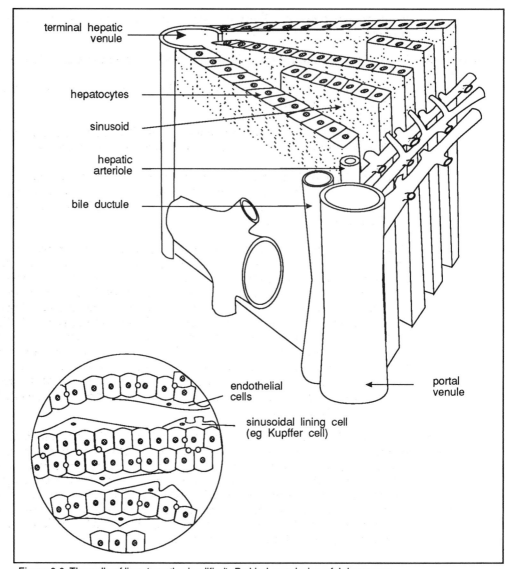

terminal hepatic
venule

hepatocytes

sinusoid

hepatic
arteriole

bile ductule

endothelial
cells

sinusoidal lining cell
(eg Kupffer cell)

portal
venule

Figure 9.6 The cells of liver (greatly simplified). By kind permission of J. Lawrence.

9.4.2 Renewal by stem cells

terminal differentiation

One of the associated properties that often accompanies terminal differentiation is that cells are incapable of cell division after specialisation. In many instances, this is the result of structural changes that occur during cell differentiation. For example, mammalian reticulocytes lose their nuclei as they terminally differentiate into erythrocytes and hence are not able to replicate. The same is true for skin cells. As skin cells move to the outer extremities they synthesise a succession of different types of keratin until eventually their nuclei degenerate producing an outer layer of dead, keratinised cells which are eventually shed. Clearly a new strategy is required in order to replace cells incapable of cell division. In such cases, cell renewal is by expansion of a specialised population of cells known as stem cells. The essential property of stem cells is that they possess the ability to divide without limit and to give rise to differentiated progeny.

properties of stem cells

More specifically, all stem cells possess the following three essential features:

- they are not terminally differentiated cells, that is they are not at the end of a differentiation pathway;

- they appear to be able to divide without limit and so, within the context of the life time of an organism, they can be considered as being immortal;

- upon division of a stem cell each daughter cell makes one of two choices: it can retain the characteristic properties of a stem cell like its parent, or it can be induced to embark upon a new pathway which will eventually lead to it becoming terminally differentiated.

Obviously, the decision between remaining a stem cell and entering a new pathway of cell differentiation is crucial and, as we shall see shortly, it is controlled by external influences. One important feature of a stem cell is that it is not intended to carry out the differentiated function, it simply serves as a source of cells that can, under appropriate stimulation, differentiate. You may at this point have the mistaken impression that all stem cells are alike and simply represent totally undifferentiated cells. That is certainly incorrect. Although stem cells may appear to have no visible signs of differentiation, their existence is the result of limited differentiation. There are different classes of stem cell which have very limited capacity to terminally differentiate. To take you back to our discussion earlier in the chapter, stem cells represent a state of determination; upon receipt of an appropriate environmental cue they express this determined state and terminally differentiate.

unipotent stem cell

pluripotent stem cell

All stem cells have a severely restricted potential to differentiate. Some can only give rise to a single differentiated cell type. Such stem cells are known as unipotent stem cells (we shall see below that the stem cell population that is responsible for regenerating skin cells belongs to this particular class). Others have a slightly broader range of possibilities and can give rise to a number of differentiated cell types. From our earlier description of stem cells, we might call these pluripotent stem cells (we shall see that the stem cell population responsible for regenerating blood cells has the ability to differentiate into all of the mature cell types found in blood).

SAQ 9.3

Differentiated cells can be renewed by either simple cell division or by the replication and differentiation of stem cells. What do you think is the most important factor that determines which of the two routes is taken for the renewal of different cell types?

SAQ 9.4	Below are a number of tissue types. Indicate in the space opposite each, whether simple cell division or stem cell differentiation is used to renew cells of the tissue. To answer this question, you must have some knowledge of the tissues described. Most of them have been covered in the text but watch out for the trick example.

1) Liver

2) Blood

3) Muscle

4) Skin

5) Nerve

9.4.3 Regeneration of skin cells

The epidermal layer of the skin and the epithelial lining of the digestive tract both contact the outside environment and are therefore subject to the most damaging encounters. As a consequence, there is a continual, major requirement for cell replacement. Damaged cells are lost from the surfaces that contact the outside environment and need to be replaced by proliferation of cells from deep within the respective layers. We shall consider the structure and replenishment of the epidermis in some detail.

proliferative unit, dermis, basal cell, basal lamina

A section through the epidermis reveals a relatively ordered structure and it is now known that the epidermis is organised into a number of proliferative units (Figure 9.7). The epidermis sits above the connective tissue of the dermis. The inner most layer consists of basal cells that are in intimate contact with the basal lamina that separates the epidermis from the dermis.

prickle cells

Lying above these basal cells are several layers of larger cells that are known as prickle cells because of their appearance under the light microscope. The outer layers of the epidermis are composed of cells which have lost all intracellular organelles and are essentially full of keratin. These outer cells are called squames. Dead cells are lost from the external surface of this layer of squames. The arrangement of these cells throughout the epidermis is thought to be quite regular. Although not always apparent, it is clear from relatively thin sections that the various cells are arranged in regular columns. This particular columnar organisation is believed to be present throughout the entire epidermis. The diameter of the column is such that about 10 basal cells form the foundations on which the column stands. Each of these columns is termed an epidermal proliferative unit.

squames

The picture that emerges about the life of one of these columns is as follows. Each column is thought to have a central basal cell that acts as a stem cell for that particular column. As mentioned above, this particular stem cell is immortal within the life span of the organism. This stem cell divides to give two daughter cells, one of which maintains the mantle of immortality. The other daughter cell, after a few further divisions, passes to the periphery of the basal cell mass at the foot of the column. It then begins to move up the column and, as it does, it differentiates to take on the appearance of the mature epidermal cells. Clearly, the crucial decision that has to be made occurs at the point of the first division of the stem cell.

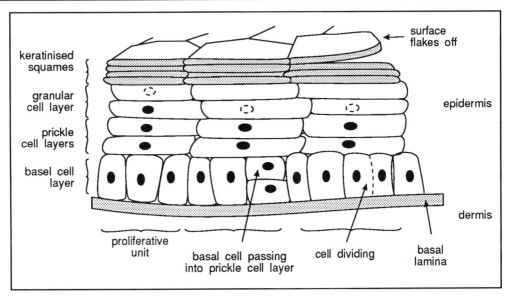

Figure 9.7 Epidermal proliferative unit.

∏ Why does one of the resulting daughters retain the characteristics of a stem cell whilst the other is dispatched to a pathway that will lead to differentiation into a keratinised squamous cell and eventual cell death? Can you think of two general mechanisms whereby such a decision could be determined? (Think of the environment of the cells and the two daughter cells produced by division of the stem cell - are they identical)?

proliferation of stem cells

In principle the fate of the daughter cells could rest on an asymmetry in the stem cell division such that the two daughter cells are not identical; one could solely inherit those characteristics that render it immortal. The second possibility is that the fate could be determined by some kind of external influence. The first possibility seems to be untenable. This can be demonstrated by studying what happens when a patch of skin is somehow lost. You know what the outcome is. The missing patch of skin is replaced by the proliferation and migration of surrounding tissue. Thus new epidermal proliferative units are formed from existing units. These new units are indistinguishable from those that have generated them and certainly possess their own stem cells. Thus during this process the number of immortal stem cells has increased. This is inconsistent with the notion that stem cell division is asymmetric.

importance of contact with the basal lamina

So, if stem cell division is not asymmetric we have to assume that the different fates of the two daughter cells following stem cell division is determined by some external influence. But, what can it be? A number of experimental results now point to the fact that the fate of the cells following stem cell division is determined by whether they remain in contact with the basal lamina. Remember, the stem cell is situated in the basal layer of cells which is in intimate contact with the basal lamina. It appears that the stem cell character is maintained as long as the cell remains in contact with the basal lamina and that changes which will eventually lead to terminal differentiation are initiated as soon as the cell loses this contact.

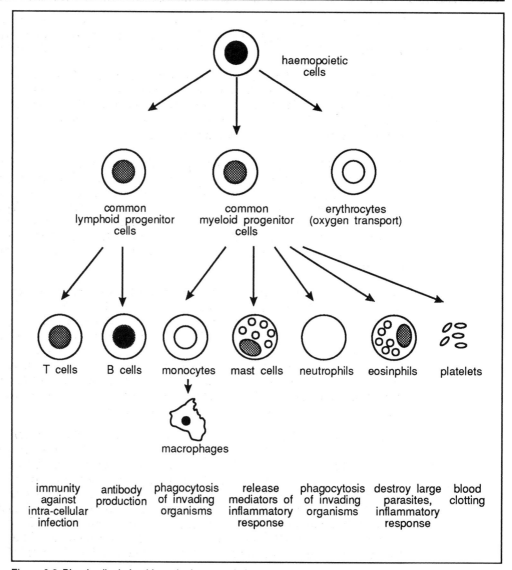

Figure 9.8 Blood cells derived from the haemopoietic stem cell.

9.4.4 Regeneration of blood cells

haemopoietic
stem cell

The blood contains many different types of differentiated cells each performing very different functions. Figure 9.8 illustrate some of these - bear in mind that this has been greatly simplified. All of them have a limited life span within the circulation and, remarkably, all are regenerated from the same stem cell population. This haemopoietic (or haematopoietic) stem cell population is therefore pluripotent.

SAQ 9.5

Using Figure 9.8, name the cells which show:

1) totipotency;

2) pluripotency.

haemopoiesis

The different kinds of blood cells require very different regeneration rates and thus regeneration must be tightly controlled in order to prevent an imbalance of cell types. This process of blood cell regeneration is known as haemopoiesis (or haematopoiesis).

bone marrow

Actually studying haematopoiesis is quite difficult. There is no regular spatial organisation, as in the epidermis, instead the production of new blood cell types takes place in the disordered confines of the bone marrow, where very many different cell types are intermingled. As a consequence, our view of precisely how blood cell types are regenerated remains somewhat incomplete.

The disorderly arrangement of cell types in the marrow means that identification of all but the terminal differentiated stages of the different cell types is not possible. The largely undifferentiated precursors of say lymphocytes and erythrocytes are indistinguishable to the eye and it certainly is not possible to identify the pluripotent stem cell that is the founder of all of the blood cell types. Fortunately, experimental systems have become available which have led to the realisation that there is a single stem cell which has the ability to give rise to all of the blood cell types following its proliferation.

irradiation

Much of the study into blood cell differentiation has used experimental animals, usually mice, that have been exposed to a large dose of radiation. We are all familiar to some degree with the potentially lethal effects of high doses of radiation. Essentially, the radiation dose, if large enough, will halt all cell division in the affected animal. The consequences are that the animal will die within a few days due to the inability to regenerate cell types, particularly blood cell types. However, the irradiated animal can be saved from death by the infusion of blood cells from a healthy donor. Among these donor cells are some which are evidently able to colonise the irradiated mouse and replenish the haematopoietic tissue lost during irradiation.

colony forming unit (CFU)

It has been established that this recolonisation is due to clones of cells, a clone being a population of cells that have originated from the proliferation of a single cell. The founder of such a clone is called a colony forming unit (CFU). Some, if not all, of the CFUs must be stem cells because it can be observed over a period of time that some of the clonal colonies are able to reproduce themselves whilst at the same time providing cells that differentiate into the various mature blood cell types. It is now possible to isolate CFU stem cells from bone marrow of normal animals and demonstrate that they are able to produce all of the mature blood types in irradiated animals, demonstrating that the stem cell population responsible for regenerating blood cells is pluripotent.

The pluripotent blood stem cell operates in exactly the same way as the epithelial stem cells described above. Division of the stem cell gives rise to new stem cells which are immortal and a population of cells that can undergo differentiation. Once a cell has differentiated as, say, an erythrocyte or a lymphocyte there is no return, that decision is not reversible. This suggests that somewhere along their pathway of differentiation, cells become committed to a particular line of differentiation. You will notice from Figure 9.8 that there is a series of stages of commitment. Thus a haemopoietic cell can become one of many types of cells, the common myeloid progenitor cells are more restricted and the mast cells show specific and terminal differentiation. This happens long before the final few divisions when the cell becomes terminally differentiated. It therefore appears that commitment to a particular line of differentiation is followed by a series of cell divisions which amplifies the number of cells of a given differentiated type. It is at this amplification step that the controls operate to coordinate the production of the specialised cell types. The controls operate to regulate the production of the different cell types according to the precise need at any one time.

9.5 Problems of cell distribution

9.5.1 Territorial stability in the adult body

compartment-
alisation

One of the most amazing aspects of the maintenance of tissues and organs, given the amount of cell regeneration that occurs in an adult organism, is that different cell types do not become jumbled and thus destroy the obvious compartmentalisation. The growth and renewal of many of the soft parts of the body is tightly regulated so that each component is adjusted to a particular niche. The restraints that prevent disorder ensuing are of various kinds. Some specialised cells are contained within tough capsules of connective tissue which physically prevent excessive spread of the cells. Some types of cells will only survive within their own environment and will die if deprived of essential environmental factors. However, perhaps the most important strategy for keeping cells in their appropriate place within the larger structure of an organism is the principle of selective adhesion, that is cells of the same type tend to stick together either in solid masses or in epithelial sheets. We will discuss this in more detail below.

Iselective
adhesionselectiv
e
adhesion

epithelial
sheets

We have already briefly discussed the importance of epithelia when we discussed *Hydra*. It is very clear that epithelial organisation helps to keep specialised cells in their correct territories. Epithelial cells are held together by their attachments to each other and to the basal lamina. The basal lamina strictly marks the boundary between the epithelium and the tissues beneath it and very few cell types can penetrate that boundary. The very few exceptions to this are lymphocytes, macrophages and nerve cell processes, all of which can invade the basal lamina. Epithelial sheets can also be used to enclose and contain other cells. For example, sheets of endothelial cells line the blood vessels and contain the blood cells, and the epithelial lining of the gut prevents the underlying connective tissue of the gut wall from invading the lumen of the gut. Thus the compartmentalisation of the body created by epithelia plays an important role in keeping cells properly segregated and confined to their correct territories.

9.5.2 Cell-to-cell adhesion and the extracellular matrix

As we have mentioned earlier, one of the great driving forces for the evolution of multicellularity was the freedom that it gives cells to specialise in ways that would not be viable as isolated cells. As we have seen during the discussion of the skin, this specialisation can even result in the death of epithelial cells so that they can provide a tough protective coat around the organism. Specialised cells in multicellular animals are normally collected into tissues where they can co-operate to perform a particular function. Different kinds of tissues can then collect to form an organ.

extracellular
matrix

Tissues represent well organised territories of specialised cell types and we can see that all the cells within a tissue are in contact with a complex mixture of extracellular proteins that is known as the extracellular matrix. The extracellular matrix not only acts as a kind of biological 'cement', designed to hold the cells of the tissue together, but also provides a medium through which cells can migrate and interact with one another. The presence of the extracellular matrix is therefore crucial for the structural integrity and function of the tissue. Equally important from the structural point of view, cells within a tissue are commonly linked to neighbours via tight cell-to-cell contacts. The points of contact between adjacent cell occur at special regions of the plasma membrane known as cell junctions. We briefly discussed communication between cells through such junctions in Chapter 3. We will be looking at them in detail later in this chapter. Many of these junctions simply serve to hold cells together, thereby contributing to the structural integrity of the tissue. Other types actually allow small molecules to pass from

cell junctions

the inside of one cell to the inside of an adjacent cell. This neighbourly cross-talk may serve to co-ordinate the activities of individual cells within the tissue and will clearly be crucial for the correct functioning of the tissue. The remaining part of this chapter will be devoted to discussing these two very important aspects.

9.5.3 Intercellular recognition and cell adhesion

There are two very general ways in which specialised cells become organised into tissues. The first, and simplest, is the formation of a tissue by the proliferation of a special founder cell whose progeny are prevented from migrating away by molecules of the extracellular matrix and by the fact that the cells adhere to each other by way of specialised cell junctions.

∏ Bearing in mind what we have already discussed, can you think of an example of a tissue that might be formed by such a mechanism?

The formation of epithelial sheets is one obvious example of this particular strategy.

∏ Can you think of a second way in which a complex tissue could be formed from its individual cells?

cell migration and sorting out

The second method is far more complex and involves the migration of specialised cells from distant locations within the organism and their subsequent association with different cell types in the correct arrangement to form a tissue. We could perhaps call this cell sorting out. Many examples of this kind of mechanism can be seen to act during embryogenesis (development of the shape and form of an embryo). This particular mechanism of tissue formation clearly requires a well organised pattern of cell movements to ensure that the correct cells end up in the correct location.

chemotaxis

contact guidance

This can be achieved by a chemical signal which attracts appropriate cell types and causes them to migrate toward the signal source, a process known as chemotaxis. It can also be achieved by laying down special pathways in the extracellular matrix akin to the railway tracks that direct the movement of trains. These pathways would then direct the migration of cells to their final destination, a process known as contact guidance. Inherent in both these mechanism is the assumption that cells must be able to recognise one another and stay together in a tissue which is distinct from surrounding cells. We actually know very little about the mechanisms that govern the ordered migration of cells and their association into tissues and organs in animals, but we may be able to get some clues by studying less complicated cells that exhibit similar, if not identical, phenomena.

9.5.4 Cellular slime moulds

There are several species of cellular slime moulds, but one in particular, *Dictyostelium*, has proved a useful model system. *Dictyostelium* is an organism that inhabits the floors of temperate forests where it feeds on the bacteria that are growing on the rotting vegetation. It has two distinct phases to its life cycle: a vegetative phase during which the organism exists as individual cells that grow and divide by simple mitosis and a spore-forming phase (Figure 9.9).

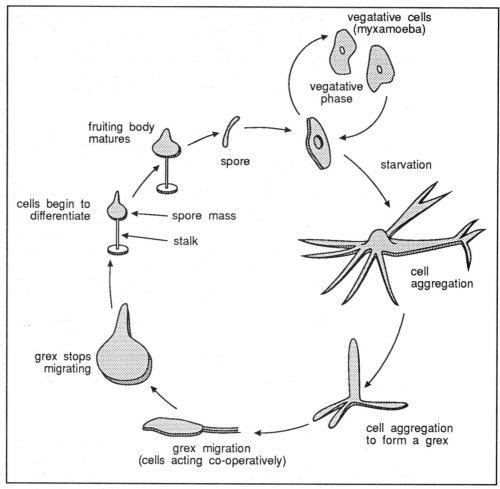

Figure 9.9 The life cycle of *Dictyostelium* (stylised).

The vegetative phase is maintained as long as the individual cells can munch away at the bacteria. However, once the food supply runs out this sets alarm bells ringing and the organism reacts by setting in train a process which will ensure its survival. The process results in the formation of spores which can survive the conditions of starvation and which can be relocated to fresh environments where the food is plentiful and where the spores can germinate to produce new cells. The part of the life cycle dedicated to spore formation is called the developmental phase. The developmental phase is initiated by starvation, although the precise mechanism whereby this occurs still remains obscure. At the very beginning of the developmental phase the cells are still individual. However, within a few hours of the onset of starvation there is an orderly movement of cells towards special centres, a process called aggregation, where the cells eventually collect into multicellular masses containing of the order of 100 000 cells. The analogy with the system of tissue formation mentioned above should be obvious.

developmental
phase

aggregation

We actually know quite a lot about precisely how aggregation is achieved in *Dictyostelium*. The centre to which the aggregating cells migrate is either a single cell, or group of cells, which acquire the ability to secrete a special chemical (3′,5′-cyclic-AMP) to which cells in the immediate vicinity can respond. The responding cells are able to

cyclic AMP

recognise the secreted chemical because they possess special receptor proteins for the chemical in their plasma membranes. Once a responding cell has bound the secreted chemical via these specific surface receptors, it is induced to migrate (chemotact) toward the source of the signal. In this way, the cells in the immediate vicinity of one of these centres collect together in a multicellular mass. The cells also possess other cell-surface molecules which allow the cells to stick together tightly and stabilise the multicellular structure.

Once formed, the cells within the multicellular mass undergo both cell differentiation, which results in the generation of two new cell populations (stalk cells and spore cells), and the original disorganised cell mass transforms so that a stalk composed of stalk cells is formed which holds aloft a spore head containing the spores. The stalk cells are dead, but the spores remain viable. The mature structure, known as a fruiting body, is such that the spores are held aloft which presumably maximises their chances of being dispersed to new locations where the food supply will be more plentiful. Clearly the biology of this rather lowly eukaryotic species offers some insight into how processes like chemotaxis and cell-to-cell adhesion might occur in the generation of tissues and organs in much more complex systems. [One of the attractions of systems like *Dictyostelium* is that aggregation takes place over a relatively large area (one aggregation centre controls a territory about 1cm in diameter) and can easily be followed on an agar surface]. It is therefore relatively easy to eavesdrop on the cellular cross-talk that takes place. These advantages are not offered by cells in a developing embryo where they communicate over microscopically small distances and often are buried within the embryonic cell mass. Incidentally, the rather simple binary choice of cell differentiation (shall I be a stalk cell or shall I be a spore?) has also proved an attractive system for researchers seeking to gain insight into how cell differentiation occurs in more complex systems.

9.5.4 Cell aggregation in sponges

Although a number of cell-surface proteins have been identified in *Dictyostelium* that might be involved in cell-to-cell adhesion, the precise molecular mechanism still eludes research workers. Fortunately, there are other lower eukaryotic systems where the molecular basis of cell-to-cell adhesion is better understood. One such system is another relatively simple organism, the sponge. Sponges are amongst the simplest of all multicellular animals consisting only of five or six cell types which can easily be dissociated into single cells by passing them through a very fine sieve. If the dissociated cells are then mixed together, they reaggregate and rearrange to eventually produce a normal sponge (Figure 9.10). Furthermore, if cells from two different species of sponge are mixed together, cells will only associate with cells from their own species.

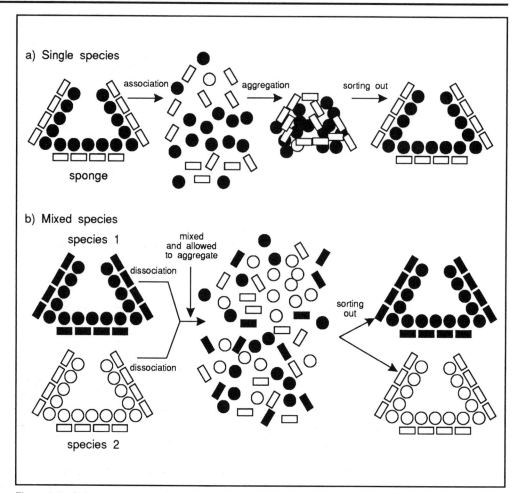

Figure 9.10 Cell re-arrangement in sponges.

aggregation factor

proteoglycans

glycoproteins

baseplates

This species-specific reaggregation is mediated by a large extracellular aggregation factor which consists of a complex mixture of proteoglycans and glycoproteins (more about these two types of molecules a little later on). The extracellular factor is actually quite large (of the order of 100nm in diameter) and is able to cause cells to adhere because it interacts with surface receptors, called baseplates, which are present on the surface of the sponge cells. Thus the mechanism for cell adhesion in sponges is one in which an extracellular matrix binds to the same protein on different cells causing them to stick together. We can represent this in diagrammatic form as in Figure 9.11.

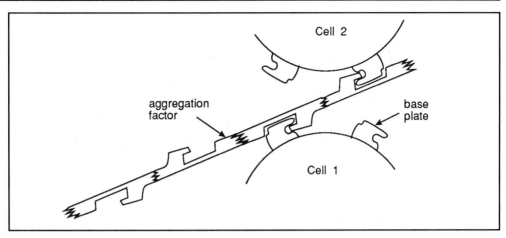

Figure 9.11 Cell-cell adhesion in sponges via baseplates and aggregation factor.

9.5.5 Vertebrate cell adhesion

The cells of adult vertebrate tissue are very difficult to dissociate, but those of the embryo can be dissociated with relative ease using proteolytic enzymes. If dissociated cells from two different embryonic tissues are mixed together they initially interact to form mixed aggregates in which cells from both tissues are randomly distributed throughout the new aggregate. However, the cells within these mixed aggregates will eventually sort themselves out to produce two distinct regions containing cells from each of the two original tissues. This is reminiscent of the situation described in sponges (Figure 9.10). With vertebrates there is a difference. If cells from two species are mixed they tend to ignore the species difference and sort out according to their tissue of origin containing cells from both species (Figure 9.12).

The pattern of redistribution that accompanies the sorting out of cells in mixed tissue aggregates usually results in the same outcome: the cells from one tissue usually end up in the centre surrounded by cells from the second tissue.

∏ Can you offer any explanation as to why cells should end up segregated into a central and outer region when they have been dissociated and allowed to re-associate? This is quite a difficult question to answer. Think about the strength of adhesion between cells. Will it be different for different cells? Now can you answer the question?

differential adherence

The accepted explanation is that cells from different tissues show differential adhesiveness. In other words, cells of different tissues show a hierarchy of adhesiveness. Thus in the example above, the cells at the centre of the mass will adhere so strongly to their own kind that they will exclude the other cell type which will be forced to the outside where they will interact with each other to form a new layer of cells. This hypothesis is known as the differential adherence hypothesis.

hypothesis

We conclude that the correct movement of cells in developing systems may be brought about by at leat two mechanisms. One involves chemotaxis and the other selective adhesion.

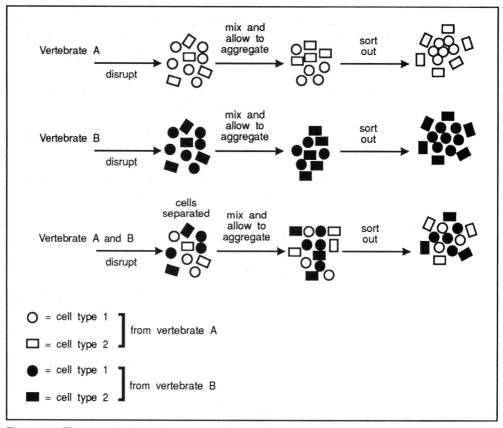

Figure 9.12 Tissue sorting in vertebrates.

9.6 Cell adhesion in tissues and organs

9.6.1 Cell junctions

Cell junctions are usually grouped into three functional groups:

- adhering junctions (desmosomes);

- impermeable junctions (tight junctions);

- communicating junctions (gap junctions, chemical synapses). Within these three general groups there are specific types of structures. We will examine some of these in a little more detail.

9.6.2 Desmosomes

Desmosomes are widely distributed in tissues where they anchor cells together enabling them to behave as a structural unit. They are very abundant in tissues which are subject to severe physical stress, for example, cardiac muscle, skin epithelium and the neck of the uterus. There are three kinds of desmosome:

- belt desmosomes - These form a continuous ring around each of the interacting cells, usually toward the apical end of the cell. The belt desmosomes in adjacent cells interact via a rather ill-defined extracellular, fibrous material which presumably holds the two interacting membranes together (Figure 9.13);

- spot desmosomes - Spot desmosomes act like cellular rivets to hold cells together at specific points of contact (Figure 9.13);

- hemidesmosomes. These can be considered as a special version of spot desmosomes which serve to anchor the basal surface of epithelial cells to the underlying basal lamina (Figure 9.13).

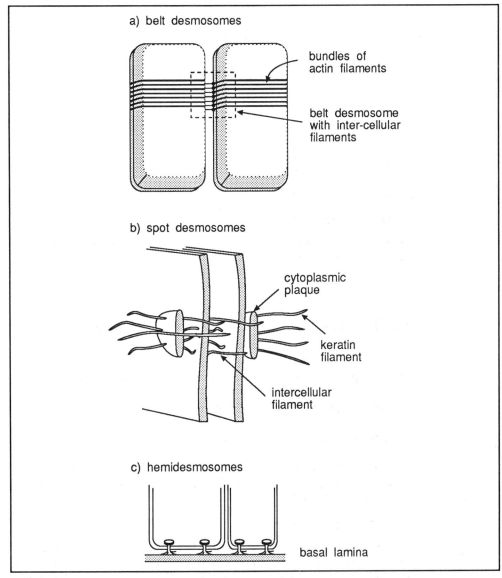

Figure 9.13 Desmosomes (stylised).

9.6.3 Tight junctions

As we have seen already, epithelial cell sheets line the surface of the body and all of its cavities where they have one very important function, namely to maintain selective permeability barriers separating the fluids inside and outside. Clearly, the junctions that hold these epithelial cells together need to maintain this impermeability preventing 'leakage' through the junction. Tight junctions play a crucial role in holding epithelial sheets together and maintaining selective permeability. The important role played by these tight junctions can be exemplified by discussing the epithelial cells that line the gut (Figure 9.14).

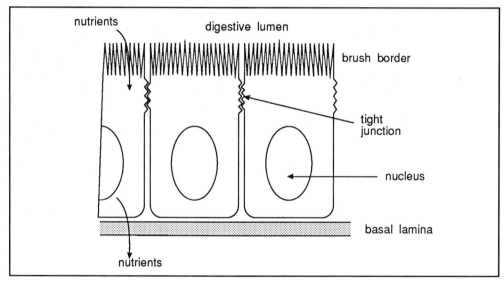

Figure 9.14 Tight junction in gut epithelium.

These cells have to be able to keep out most of the contents of the gut in the lumen whilst selectively transporting nutrients across the epithelial sheet to the extracellular fluid on the other side from where they are absorbed into the blood system for delivery to the tissues. This unidirectional transport of nutrients relies on two special pumping systems. One is specifically localised on the apical surface of the epithelial cell (that is the surface that is in immediate contact with the contents of the gut) where it can pump useful nutrients into the epithelial cells. The other is found in the basal surface and pumps these nutrients out of the cell and into the extracellular space. These two biological pumps are quite distinct and are composed of different membrane transport proteins.

∏ Can you think of what role tight junctions play in this process of unidirectional transport (Look again at Figure 9.14 and think of the movement of structures *within* rather than *through* the membrane)?

In fact they have two very important functions. It is essential that the two pumps do not diffuse through what is in reality a continuous plasma membrane surrounding the epithelial cell: The apical pump must be confined to the plasma membrane region that contacts the lumen of the gut, whilst the basal pumps must remain in the plasma membrane at the bottom of the epithelial cell. Tight junctions act as diffusion barriers within the plasma membrane and thus ensure that the apical and basal pumps are retained in their appropriate positions. It is also essential that once nutrients have crossed the epithelial cell that they do not diffuse back into the lumen of the gut through the space that inevitably occurs between adjacent epithelial cells. Again the tight junction prevents this unwanted reverse movement of nutrients by creating impermeable seals between adjacent epithelial cells to create a continuous sheet of cells.

9.6.4 Gap junctions

We dealt with gap junctions in Chapter 3. We remind you that gap junctions are the commonest type of cell junction and actually connect the cytoplasm in adjacent cells. This physical connection allows small molecules to diffuse from the cytoplasm of one cell to another. A gap junction could be considered as a very small pore. The diameter of the pore is such that relatively small molecules (inorganic ions, sugars, nucleotides, amino acids and vitamins) can freely diffuse through whilst movement of larger molecules (proteins, nucleic acids and polysaccharides) is absolutely restricted. Gap junctions are composed of proteins which extend out from the plasma membrane of adjacent cells. The protein connection forms a structure known as a connexon which provides an aqueous channel between the two adjacent cells. Each cell provides half of the proteins that are required to form a connexon. The connexons join the plasma membrane of adjacent cells to leave a gap which is approximately 2 to 4nm wide.

SAQ 9.6	1) In which tissue types will there be many desmosomes? Name three examples.
	2) Explain the function of tight junctions in the intestinal epithelium.

9.7 The extracellular matrix

In animals, the extracellular matrix is a mixture of protein and polysaccharide molecules that are elaborated and secreted locally by cells to form a well organised mesh in which the cells of a tissue are embedded. In other words, the extracellular matrix acts as a universal biological 'cement' to hold cells together within a tissue. However, this is a great oversimplification and it is now clear that the extracellular matrix does a great deal more than just bind cells together within a tissue. It also plays a complex role in regulating the behaviour of cells that contact it.

collagens

glycosamino-glycans

proteoglycans

The macromolecules that make up the extracellular matrix are secreted locally by cells, especially fibroblasts which are widely distributed throughout the matrix. The matrix has two major molecular components: the collagens and the polysaccharide glycosaminoglycans, which are normally covalently attached to a protein to give proteoglycans. The polysaccharide components form a kind of a gel in which the collagen fibres are embedded. The collagen fibres give the matrix strength and help with its organisation, whilst the aqueous phase of the polysaccharide gel permits the diffusion of molecules through the matrix.

elastin, fibronectin, laminin

Three other proteins are also frequently found in the extracellular matrix of certain tissues: elastin which imparts a degree of elasticity to the matrix in which it is embedded, and two glycoproteins called fibronectin and laminin.

∏   What do you think the term connective tissue means?

It is the term used to describe the extracellular matrix and the cells that are found in it (chiefly fibroblasts, macrophages and mast cells). The quantity of connective tissue and its composition depend to a large extent on the organ. Bone and cartilage are rich in connective tissue whereas brain and the spinal cord contain relatively little of it.

9.7.1 Collagen

tropocollagen

collagen fibrils and fibres

Collagen represents one of the major proteins found in mammals, amounting to 25% of all mammalian protein. Because of its special amino acid composition collagen forms long fibres which have an enormous tensile strength. It consists of three polypeptide chains, each containing approximately 1000 amino acids, which have an helical arrangement and which are wrapped around each other to from a rope-like molecule called tropocollagen. Interaction of five tropocollagen molecules leads to the formation of microfibrils which in turn associate into collagen fibrils. A number of collagen fibrils interact to form a collagen fibre (Figure 9.15).

∏ Use Figure 9.15 to help you calculate how many peptide chains are present in a collagen fibre composed of 20 collagen fibrils.

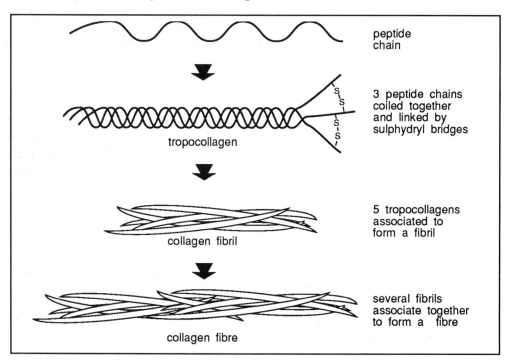

Figure 9.15 Collagen fibres.

Your calculation should have revealed that such a fibre would contain 300 (20 x 5 x 3) peptide chains. The composition of the polypeptide chains depends on the type of collagen. There are at least five different types of collagen, each of which has its own characteristic properties. Hence they occur in different tissues and in different proportions (Table 9.1). Up to 90% of the collagen of mammals is of Type 1.

Tissue	
Skin, tendon, bone, ligaments, cornea, internal organs	Type 1
Cartilage, fluid of the eye	Type 2
Skin, blood vessels	Type 3
Basal lamina of epithelia	Type 4

Table 9.1. The occurrence of different types of collagen. Type 5 is only present in small amounts but is widely distributed.

33% glycine

proline

post-translational hydroxylation

glycosylation

The polypeptide chains in tropocollagen show a specific, quite distinctive composition and sequence of amino acids. Every third amino acid in the polypeptide is a glycine residue and the amino acid proline is also present in larger quantities than in most other proteins. In addition, collagen is rich in two unusual amino acids, hydroxyproline and hydroxylysine. Both hydroxyproline and hydroxylysine arise as a result of the post-translational hydroxylation of proline and lysine, respectively. By post-translational hydroxylation we mean that the hydroxyl groups are added after the proline and lysine have been incorporated into the protein. These hydroxylations occur before the polypeptide chains have associated to produce tropocollagen. Several of the hydroxylysine residues become further modified by the addition of carbohydrate, a process known as glycosylation. Glycosylation of hydroxylysine usually involves the addition of a galactose-glucose unit.

The polypeptide chains that are eventually going to make up collagen are synthesised and secreted by fibroblasts. Following their synthesis, three polypeptides associate to form tropocollagen which is then secreted. Once the tropocollagen molecules enter the extracellular space they interact with each other first to form fibrils and then fibrils interact to form fibres. We might imagine that they are held together by hydrogen bonds.

∏ How could the interaction between the fibrils and fibres be strengthened?

chemical cross-linking

The answer is by the formation of chemical cross-links between residues within polypeptides making up the fibre (eg -S-S-bridges). The degree of cross-linking critically determines the tensile strength of the resultant collagen, for example, there is a high degree of cross-linking of the collagen that makes up the Achilles tendon where tensile strength is crucial.

Collagen secreting cells are able to adapt the collagen component of the extracellular matrix to suit the needs of the tissue. They can synthesise different forms by incorporating the different types of basic polypeptide into the tropocollagen molecule. Further variation can be achieved by different degrees of hydroxylation and glycosylation. Finally, following secretion, the tropocollagen assembles into fibrils which can be cross-linked to a greater or lesser degree

9.7.2 Elastin

Elastin is an important component of connective tissue, particularly of the more elastic connective tissue which occurs in the lungs, the walls of blood vessels and ligaments. Like collagen, elastin has a very distinctive amino acid composition. About 30% of its amino acids are glycine and there is also a considerable amount of proline. Hydroxylated amino acids (hydroxyproline and hydroxylysine) are rare, as are other polar amino acids. However, characteristically, elastin has a high concentration of hydrophobic amino acids. Elastin molecules are secreted into the extracellular space where they form filaments and sheets in which the elastin molecules are cross-linked to each other to create an extensive network. Unusually for proteins, the polypeptide backbone of the elastin molecules remains largely unfolded in a structure known as a random coil. The high concentration of hydrophobic amino acid residues is largely responsible for the random coil structure of elastin. It is the cross-linked, random coiled structure of the elastin fibre network that allows it to stretch and gives it its elasticity (see Figure 9.16).

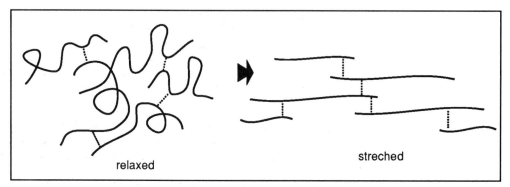

relaxed

streched

Figure 9.16 Elastic network formed by means of intermolecular links between the random coils of elastin.

9.7.3 Fibronectin

adhesive
protein

Fibronectin is an important protein constituent of the extracellular matrix where it is involved in cell-to-cell adhesion and in binding cells to substrata such as the basal lamina. It is therefore no surprise that it is called an adhesive protein. It is found at the surface of nearly every cell. The majority of the fibronectin in the extracellular matrix occurs in the form of large aggregates. It is a glycoprotein which consists of two subunits that are held together by disulphide bridges.

9.7.4 Glycosaminoglycans

N-acetyl-
glucosamine

N-acetyl-
galactosamine

These are molecules which are formed by a group of polysaccharide chains which consist of repeating units of disaccharides. One of the two sugars in the disaccharide repeat is always either N-acetylglucosamine or N-acetylgalactosamine. The glycosaminoglycans have a strong polyanionic (negatively charged) character which results from the many sulphate and carboxyl groups that are present in the sugar residues. Glycosaminoglycans can be categorised into six groups according to the composition of the repeating disaccharide units (Table 9.2).

⫟ In what way does hyaluronic acid differ from the other glucosaminoglycans listed in Table 9.2 and which component of the repeating disaccharides carries a negative charge?

From the table you should have spotted that hyaluronic acid does not contain sulphate groups. It also differs from all the other glycosaminoglycans in one important respect, it is *not* attached to proteins. A common component of the repeating disaccharide of glycosaminoglycans which carries a negative charge is D-glucuronic acid.

All glycosaminoglycans, other than hyaluronic acid, are attached to proteins to give proteoglycans. The polysaccharide moieties are attached to the protein, known as the core protein, via serine residues in the amino acid backbone.

Glycosamine	Repeating disaccharide	Sulphate residue/unit
Hyaluronic acid	D-glucuronic acid - N-acetylglucosamine	No
Chondroitin sulphate	D-glucuronic acid - N-acetylgalactosamine	Yes
Dermatin sulphate	D-glucuronic acid - N-acetylgalactosamine	Yes
Heparin sulphate	D-glucuronic acid - N-acetylglucosamine	Yes
Keratin	D-galactose - N-acetylglucosamine	Yes

Table 9.2 The common glycosaminoglycans.

The variety in proteoglycan structure is very great indeed since the type of protein as well as the extent of sulphation may vary considerably. The macromolecular structure of proteoglycans is often highly organised, as exemplified by the structure of the proteoglycan found in cartilage. This kind of proteoglycan exists in the form of large aggregates which have molecular weights in excess of 10^8 daltons. Each core protein contains approximately 1900 amino acids to which are attached about 100 chondroitin sulphate groups and about 40 keratin sulphate chains (Figure 9.17).

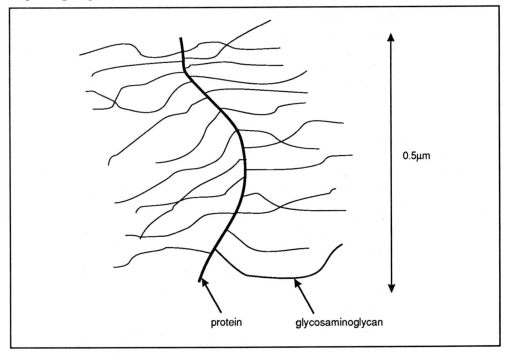

0.5μm

protein   glycosaminoglycan

Figure 9.17 A stylised proteoglycan.

hydrated gels

Glycosaminoglycans have a very large volume to mass ratio. This enables them to form strongly hydrated gels which occupy the major part of the extracellular space.

∏ What do you think are the functions of glycosaminoglycans? (Think of the gels they may produce).

One of their major functions is to provide protection against mechanical pressure. The gels provide stability to the extracellular matrix of joints and other tissues which are exposed to mechanical deformation. What you might find suprising is that proteoglycans are also involved in the regulation of tissue growth and repair and in the transport of molecules and ions through the extracellular matrix.

SAQ 9.7

What is the difference between a proteoglycan and a glycoprotein?

SAQ 9.8

Using the list provided below, identify the compounds described as:

1) a polymer containing D-glucuronic acid and N-acetylglucosamine but with no sulphate;

2) a protein rich in glycine, hydroxyproline and proline which normally consists of three peptide chains twisted together in a helix;

3) a protein which displays a considerable amount of random coiling;

4) a polymer containing D-glucuronic acid, N-acetylglucosamine and sulphate residues.

Use words from this list: heparin, elastin, laminin, hyaluronic acid, chondroitin sulphate, tropocollagen, fibronectin.

9.7.5 Basal lamina

The basal lamina is a specialised extracellular matrix which is found at the base of the epithelial cell layers of the skin and between two layers of different cell types, for example between muscle cells and the surrounding tissue. Its main constituents are type 4 collagen, proteoglycan, fibronectin and laminin. Laminin is a glycoprotein which consists of two protein subunits which are linked via disulphide bridges. The basal lamina has a number of functions.

permeability barrier

Apart from its obvious structural role, it also constitutes a permeability barrier. You may know that the kidneys filter out unwanted/waste materials from the blood. Basal lamina is involved in the filtration process that occurs in the renal glomeruli (between the renal epithelium and the vascular endothelium). It can also display selective permeability toward cells. The basal lamina that lies between epithelium and connective tissue will allow leucocytes (white blood cells) to migrate through but denies such access to fibroblasts.

∏ What else does the basal lamina do?

Earlier, when we discussed the regeneration of epithelial cells, we described the basal lamina as also being involved in inducing cell differentiation. We hope you remembered this! When the skin stem cell divides one daughter retains the mantle of immortality whilst the other is destined for terminal differentiation and ultimate cell death. Immortality in this case is retained by the cell that remains in contact with the basal lamina. The basal lamina is produced by cells which are situated on this layer.

9.8 Do multicellular plants and animals display the same properties?

Let us summarise what we have learnt about cells in multicellular organisms.

We have learnt that there is a high degree of specialisation amongst cells and that cells in multicellular systems may be rapidly turned over. Some are replaced directly others are derived from proliferative stem cells which may display totipotency, pluripotency or unipotency.

We have also learnt that for animals, cells which have been disaggregated may re-associate and that, when they do, they usually display an ability to sort themselves out into specific areas of tissues. We have seen that the cells in animals may be produced in one part of the body and migrate to another. We have also learn that the animal cells are in contact.

We are all aware of the fundamental differences between plants and animals such as the lack of motility in high plants. This is a reflection of the fundamental differences in the cells found in both groups. Those in plants have rigid cell walls and are not pliable. The whole distribution of cell types in plants is much more static. There is no migration of cells around the body. Thus although there is a considerable cellular differentiation in plants, the cells differentiate *in situ*.

meristem
We illustrate this in Figure 9.18. In this figure, we see a root cap. Just behind the root cap is a meristem (meristem = region of cell division in a plant). The daughter cells produced by the meristem are at first toti- or pluri-potent but as they age they first elongate and then differentiate. Some become cells involved in transport (xylem and phloem), others provide physical protection (epidermis) while others become absorptive (root hairs) etc.

Thus we can recognise that in plants we are still faced with the problems of cell differentiation and morphogenesis (eg why are particular cells produced in a particular place and how is their production controlled to give the right number). But we are not faced by the problems of cell sorting out or of such a complex intracellular matrix. Remember that the shape of a plant is largely determined by the shape of each plant cell and by the numbers and types of cells produced. In Chapter 4 we learnt that the shape of each plant cell is governed by the way cellulose is deposited in its cell walls which in turn is governed by the microtubules in the cell.

In contrast, the shape of an animal is, in part, governed by the final arrangement of cells but is especially dependent upon the organisation of the connective tissues and the nature of the intercellular matrix.

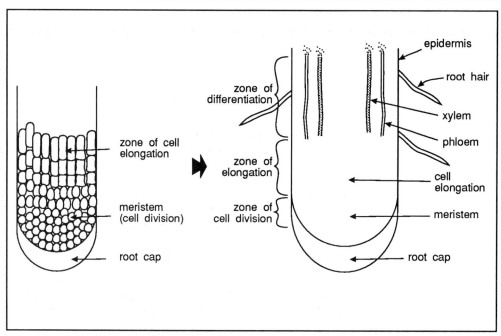

Figure 9.18 The development of a plant root.

Whether or not we focus on plants or on animals, a central issue is the need for cells to communicate with each other. Thus in plants, there needs to be co-ordination between roots and shoots whilst in animals, the activities of internal organs like the liver, heart, spleen and kidneys needs to be integrated with each other. The integration of the activities of organs is achieved through a complex cell-to-cell signalling system based upon a variety of chemical (hormone) and physical signals.

Summary and objectives

In the previous section we summarised what we had learnt about the issues arising from multicellularity. Now that you have completed this chapter you should be able to:

- explain what is meant by the terms cell differentiation, totipotency, pluripotency, determination and commitment as applied to the development of multicellular systems;

- use suitable examples to explain the role of stem cells in the replenishment of certain cell types;

- describe the physical connections that may exist between animal cells and describe functions and occurrence of these connections;

- explain the importance of the intercellular matrix in animals and be able to recognise and describe the major macromolecules that make up the matrix;

- make a comparison between plants and animals in terms of cell differentiation and the generation of shape.

Concluding Remarks

Cells are remarkable chemical factories, converting a variety of nutrients into a vast array of new and complex chemical products. Now that you have completed this volume, you can be left in no doubt that the success of cells as mediators of chemical change, is a reflection, in no small part, of the high degree of order they impose on their activities. You have learnt that cells are not merely bags or boxes of metabolic activity. Just as in the most efficient of factories, the activities in cells are tightly organized and controlled. Thus cells contain distinctive sub-cellular structures committed to carry out different activities, be it as energy generators (eg mitochondria and chloroplasts), information stores (nucleus), product manufacture (ribosomes) or product handling (eg Golgi apparatus). There is also a highly organised and regulated pathway of information from the management centre (nucleus) to the shop floor (cytosol).

Central to this organization is the acquisition and transmission of instructions stored within the cell's genetic material. The activity of the cell is not however solely dependent upon its genetic store. You have learnt, through the sections dealing with multicellularity, that cells containing the same genetic information can be quite different in appearance and activities (ie they are differentiated). Thus in multicellular systems, we find that there are a wide variety of different chemical factories, with their own special function. It is clear that the genetic information in a cell can be used in a selective manner. We can again draw the analogy with man-made factories. The activities of a factory not only depend upon its personnel and machinery, it also needs to respond to the signals it receives from its commercial, legal and social environment. Biological 'factories' also respond to the signals they receive, be they nutritional or hormonal.

Just as there are an enormous variety of factories within the manufacturing industries, in multicellular systems there is an enormous diversity of cell types. In biological terms, this great diversity of cell type is co-ordinated and controlled to produce the functional organisms. Here biological systems clearly are greater masters of their endeavours than are the efforts of humans to manage and co-ordinate their enterprises. In biotechnological terms, it should be recognized that the true potential of cells, either individually or as components of multicellular organisms, to act as highly efficient chemical factories is enormous, be they used as mediators of chemical change or as producers of valuable products.

Although this text will have provided you with a good understanding of the organisation of biological material at the cellular level, you should recognize that this is a foundation on which to build. Each of the aspects of cell biology covered in this text underpins important aspects of applied biology and biotechnology. The BIOTOL series offers you a wide variety of opportunities to build upon this foundation. It may be that you wish to learn more about the chemical changes mediated by cells (metabolism) or how to isolate and analyze the compounds produced by biological systems (analytical biochemistry) or how the expression of genetic information can be regulated or manipulated (genetics) or how the cell's function is reflected in the properties of larger organisms (physiology). The BIOTOL series is specifically designed to enable you to select those aspects in which you seek to gain further knowledge and expertise. We hope that you will grasp the opportunity to build upon the foundations you have gained.

Responses to SAQs

Responses to Chapter 1 SAQs

1.1

1) Culture 1, 6 and 8. Cocci is the term used to describe spherical shaped cells. There are also some cocci in culture 4, but these are not in predominating numbers.

2) Cultures 1, 2, 5 and 6. The arrangement of cells often give clues as to the number of planes the cells may divide in. From the drawing, it would appear that cultures 1, 2, 5, and 6 divide in one plane only, since they are in chains or stacks. The cultures which contain only single cells (ie 3, 4, and 7) provide little evidence for whether or not the cells can only divide in one plane. We have to be careful in interpreting the drawings of cultures. If the palisades of cells are only one cell thick, then these cells probably only divide in one plane. If the palisades are more than one cell thick, then they probably divide in more than one plane.

3) Culture 8. This can best be described as being composed of staphylococci. Staphylococci divide irregularly in any plane, producing irregular clusters of cocci.

4) Culture 4. Pleomorphism is the term that is used to describe cultures which contain variously shaped cells. Of course a mixed culture containing a variety of organisms many have a similar appearance.

5) Culture 3. The 'V' shaped arrangement of cells indicates that these bacilli are dividing by snapping division.

6) Vibrio is the term applied to describe cells which are shaped like commas or shortened spirals. None of the cultures illustrated contain vibrios.

7) Culture 1. Streptococci are cocci which divide in only one plane and therefore produce linear chains as seen in culture 1. Culture 6 is also comprised of cocci which appear to be dividing in only one plane so they too might be considered as streptococci. In this case, however, the chains tend to break up into double or pairs of cells, so it is perhaps better to consider these as diplococci.

8) Culture 6. The largest cells are likely to be the slowest growing. Remember that the smaller the cell, the larger the surface : volume ratio. Thus small cells have proportionally larger surface areas over which to absorb nutrients and can, therefore, support faster rates of metabolism and growth. However be careful. Not all very small cells have very fast growth rates. Some other features (eg low permeability of nutrients through the plasma membrane, low levels of enzymes) may limit the rate of metabolism.

1.2

1) The primary (first) stain used is Carbol Fuchsin.

Heat is used to 'fix' the primary stain. Carbol Fuchsin reacts more strongly with some cell constituents at elevated temperatures.

The dilute acid acts as the challenge. With some cells the dilute acid is sufficient to wash out the primary stain (ie non-acid fast cells). In other cases, the dilute acid is not able to remove the primary stain.

The Methylene Blue acts as the counter stain which stains the cells decolorised by the dilute acid.

2) The procedure is called the acid fast stain because it differentiates between cells in which the primary stain is held fast even in acid and those cells in which the acid removes the dye.

3) Non-acid fast cells lose the primary stain (Carbol Fuchsin) when challenged with acid but take up the counterstain (Methylene Blue) and, therefore, appear blue.

1.3   In our earlier calculation, we determined the limits of resolution of a light microscope was about 0.2 μm. (ie we cannot distinguish structures which are smaller than about or thinner than about 0.2 μm). The only structures included in Figure 1.3 which might be seen directly through a microscope are the endospore, the capsule and possibly the inclusion bodies. All of the other structures are too small or too thin to be seen. For example; although each flagellum is about 5 μm long it is only about 0.02 μm thick and is much too small to see directly.

What we have just demonstrated is that you should not anticipate observing many of the detailed features of prokaryotic cells even by using the most powerful light microscope.

1.4   1) Your scheme should be similar to this

2) Although dextran is made mainly of glucose, each molecule has a single fructose moiety at one end. Similarly levan is made mainly of fructose but has one glucose residue per molecule.

1.5   Gram positive 1) 3) 4) 6) 7) 10)

Gram negative 2) 4) 6) 8) 9) 10)

archaebacteria 1) 5) 10)

1) Gram positive cell walls appear more-or-less homogenous. Gram negative cell walls are composed of two layers, a thin inner layer of peptidoglycan and an outer lipid membrane. The answers to many of the remaining questions, arise from these basic facts. The cell walls of very few archaebacteria have been examined in detail, most, however, appear to be quite homogenous.

2) Gram negative cell walls have an outer membranous layer.

3) The cell walls of Gram positive cells are composed of a homogenous layer in which peptidoglycan is the major component, thus Gram positive cell walls can contain 40-90% peptidoglycan by dry weight.

4) Both Gram positive and Gram negative cells contain N-acetylmuramic acid. It is one of the central components of peptidoglycan.

5) The peptidoglycan of archaebacteria contains N-acetyltalosaminuronic acid instead of N-acetylmuramic acid.

6) Both Gram positive and Gram negative cell walls may contain some amino acids in the D configuration. The cell walls of archaebacteria only contain amino acids in the L configuration.

7) Teichoic acids, whether they contain ribitol or glycerol, are characteristic of Gram positive cell walls.

8) Gram negative cell walls contain an outer membrane composed of phospholipids and lipopolysaccharides.

9) The flagella of Gram negative cells have a basal body which has four sets of rings (L, P, S and M) while those of Gram positive cells only have two sets of rings (S and M).

10) All rigid-celled bacteria (ie Gram positive, Gram negative and archaebacteria) contain N-acetylglucosamine as a component of the peptidoglycan of their cell walls.

1.6   1) The real clue here, is the presence of the sterol, cholesterol. Sterols are not found in significant amounts in any prokaryotic cell types. Therefore they are not found in eubacteria, mycoplasmas and cyanobacteria (which are specific groups of eubacteria), nor in archaebacteria. They are however found in significant quantities in eukaryotic membranes. The cells described in this question are, therefore, eukaryotes. Note however stearic acid is common in all of the organisms listed except the archaebacteria.

2) Mesosomes are common structures in prokaryotic cells of all types and, therefore, the correct answer is all of the groups listed except the eukaryotes.

Hopanoids are polycyclic alternatives to the sterols found in eukaryotic cells. They are common in most groups of prokaryotes so again, the correct answer is all of the groups listed except the eukaryotes.

Phospholipids containing isopentanyl derivatives are quite restricted in occurrence. They are produced by archaebacteria. All other prokaryotes and eukaryotes produce phospholipids with mainly linear acyl side chains.

3) This is quite a difficult question to answer.

a) Penicillin is known to inhibit peptidoglycan synthesis. Thus in a growing culture of eubacteria, the cell walls will become stretched as the cells get bigger. As they stretch, they will become thinner and, ultimately in dilute solutions, the cells will take up water by osmosis and the plasma membrane will break (ie lysis will take place).

b) In a non-growing culture, the cells do not need to produce peptidoglycan as they are not getting any bigger. The presence of penicillin will therefore have very little effect on these cells. Penicillin only kills cells (ie will show bacterocidal activity) that are actively growing.

c) L-forms are large cells which do not have peptidoglycan in the surface layers. Thus penicillin does not have any effect on these cells (ie they are penicillin resistant). To grow, L-forms must be protected osmotically because they have no rigid cell wall and are susceptible to lysis in dilute solutions.

1.7

1) Basic dyes are dyes which carry positive charges and are therefore attracted to compounds which carry negative charges. In the list provided all except the basic proteins carry negative charges and would therefore attract basic dyes. We must however be cautious about poly β-hydroxybutyrate. Poly β-hydroxybutyrate, apart from one terminal carboxylic acid residue per molecule, is largely non-polar (ie hydrophobic) and does not readily absorb polar dyes. Basic proteins contain many amino groups which carry a positive charge ($-NH_3^+$) and therefore repel positively charged dyes.

2) Sudan Black is a fat soluble dye. Organism A apparently produces 'fat' granules when grown on medium 1 but not on medium 2. The most common 'fat' granules produced by prokaryotes are comprised of poly β-hydroxybutyrate. Thus it appears that this organism can produce poly β-hydroxybutyrate on medium 1 but not on medium 2. The likelihood is that is that medium 1 is a richer medium, containing easier to metabolise substrate than medium 2 or that medium 1 contains substantial amounts of energy rich nutrients but is short of some other essential components. In either case, it makes biological sense for the organism to store the energy-providing materials that are available.

The underlying message is that the appearance and composition of cells may, to some extent be a reflection of the environment in which they are growing.

3) This question has a similar message to that of 2). Metachromatic granules are granules of polyphosphates. It would appear that organism B can produce these granules in medium 3 but not in 4. In all probability medium 3 has an abundance of phosphate, but medium 4 has much less available phosphate.

4) The most likely explanation of the sequence of events described in the question is that the cells are capable of producing gas vesicles. Thus we begin with a suspension of cells. The sharp blow on the rubber bung cause a hydrostatic shock wave to pass through the culture causing the gas vesicles to burst. The cells, now no longer buoyed up by these little 'floatation bags', sink to the bottom of the vessel because they are denser than the medium.

After about 20 minutes, the cells begin to produce new gas-filled vesicles which cause the cells to become less dense. The cells begin to rise to the top of the vessel. Remember that air cannot be pumped back into vesicles that have collapsed. Such gas vesicles are common in cyanobacteria.

1.8

1) Pilus - predominantly thought to be involved in the attachment of cells to a substratum. Some pili however play a role in the transfer of genetic material between cells.

2) Flagellum - involved in dispersal.

3) Mesosome - plays a role in the replication of DNA and is the point at which invagination of the plasma membrane takes place during cell division. It is therefore involved in DNA synthesis.

4) 70s ribosomes - the machinery for the synthesis of proteins. Those from eukaryotes have a sedimentation coefficient of 80s.

5) Metachromatic granules - are polyphosphate granules and are involved in phosphate storage. They also might help cells to survive/grow in environments depleted of phosphates.

6) Peptidoglycan layer - present in both Gram positive and Gram negative cell walls where it provides rigidity to the structure.

7) Poly β-hydroxybutyrate granules - provide an energy and carbon reserve for cells to use in periods of starvation.

8) Endospore - enables cells to survive in hostile environments. Being very small, if they become desiccated (dry), they tend to get blown about. Therefore they aid dispersal as well.

9) Capsules - role is not always clear. In some cases they appear to reduce the rate of desiccation and in pathogens prevent the cells from being engulfed by white blood cells in animals. Perhaps we should consider them as being aids for survival in hostile environments.

Responses to Chapter 2 SAQ's

2.1   The nucleus was the first component of eukaryotic cells to be discovered. This is most likely because it is relatively large, being bigger than many prokaryotic cells, and because the DNA which it contains is easy to stain producing a strong contrast between it and the rest of the cell. Thus it can be seen using even primitive light microscopes.

2.2   3) and 4) are correct.

1) is incorrect because photosynthetic bacteria have membranes.

Some may regard 2) as incorrect because prokaryotes contain ribosomes. Some authorities, however, do not regard ribosomes as organelles. They maintain that organelles are membrane bound sub-cellular structures. It is therefore not clear cut whether or not prokaryotes contain organelles. It is certainly true however that prokaryotes do not contain the array of sub-cellular organelles displayed by eukaryotes.

2.3   1) The nucleus, mitochondria, Golgi apparatus, endoplasmic reticulum, microtubules and cytoskeletal filaments. The ribosomes are not generally referred to as organelles but it is important to note that they are present in both plant and animal cells.

2) Yes. Figure 2.3. shows that the plant cell contains a large vacuole and also chloroplasts.

3) Yes. Firstly the plant cell is enclosed within a thick coat called the cell wall. Secondly there are structures which appear to pass through the cell wall connecting the plasma membrane of adjacent cells. Finally plant cells do not have lysosomes.

2.4   1) with f) The nucleus contains the genetic material which controls cellular activities.

2) with h) Chloroplasts are the sites of photosynthesis.

3) with e) Mitochondria are the sites of aerobic respiration.

4) with b) Ribosomes are the sites of protein synthesis.

5) with g) Rough endoplasmic reticulum contains ribosomes which make proteins for export.

6) with c) The Golgi apparatus makes and exports cell wall components.

7) with d) The centriole is involved in cell division.

8) with i) Intercellular air spaces are involved in gas exchange.

9) with a) Lysosomes (in animal cells) are involved in intracellular digestion.

2.5

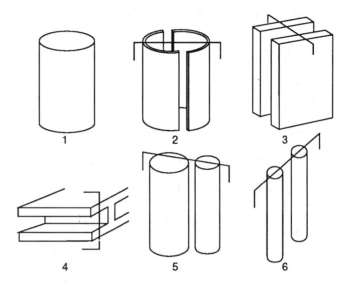

2), 3), 4), 5) and 6). The plane of section is indicated showing how these four structures could give rise to the shape showin in the question.

2.6   None of them are entirely correct. Answer 1) is correct if you exclude ribosomes as organelles. Plant vacuoles are used to store toxic compounds but most organelles do not. Chloroplasts store starch but most organelles do not. Mitochondria and chloroplasts contain all the components of particular metabolic pathways but some organelles do not.

2.7   1) Golgi apparatus.

2) Rough endoplasmic reticulum.

3) Nucleus.

4) Plasmalemma (plasma membrane).

5) Nuclear membrane.

6) Nucleolus.

Responses to Chapter 3 SAQs

3.1

1) The arrangement would be as is in a). Here the non-polar tails associate with each other in the centre of the bilayer away from the polar water of the aqueous cytoplasm and extracellular fluid.

2) Lipid in a non-polar solvent, eg hexane, would arrange itself as in b) so that the polar heads were hidden from the solvent.

3.2

Under extreme plasmolysis, the solute concentration in the protoplast can rise so high that it causes denaturation of proteins. Such proteins can no longer function and the plant may die as a result.

3.3

Magnesium and potassium ions appear to have been actively transported into the cell because of their differences in concentration on either side of the plasma membrane. Glucose is present at equal concentration on both sides of the membrane and this could theoretically be achieved by diffusion. Sodium is either prevented from entering the cell or is pumped out of the cell against the concentration gradient. Calcium is in roughly equal concentrations although most of the calcium ions inside of the cell are bound by cellular components.

3.4

The shaded portion is most likely to predominantly contain hydrophobic amino acids thus associating with the hydrophobic part of the bilayer. The non-shaded portion is likely to be polar, thus associating with charged layers of the phospholipid bilayer.

3.5

1) b)

2) Within certain limits an increase in temperature would increase the rate of migration since the migration is a kind of thermal diffusion.

3) This could be the temperature at which the membrane lipid solidifies, in which case the proteins would be held firmly in place.

3.6

In exocytosis two cytoplasm-facing sides of the membrane join first whereas in endocytosis two externally-facing sides fuse first.

3.7

Because of their potential diversity, membrane proteins might provide the recognition sites upon which fusion depends. Thus exocytosis might depend on proteins on the cytoplasm - facing side of the membrane, while endocytosis depends on the proteins on the external membrane surface. Examples of this are receptor-mediated endocytosis and exocytosis.

3.8

Theoretically budding might occur from any of the membrane systems within the cell. In actual fact it occurs mainly on the endoplasmic reticulum and the Golgi apparatus.

3.9

	Exocytosis	Endocytosis	
1)	x		
2)		x	
3)	x		
4)		x	
5)	x	x	a
6)	x	x	
7)	x	x	b
8)	x		

a) Note that not all vesicles produced by endocytosis fuse with lysosomes.

b) Entry into the cell is by endocytosis but the vesicle exits by exocytosis.

Responses to Chapter 4 SAQs

4.1 3) is correct. 5) is correct but needs elaboration. Intermediate filaments may well form a bridge between microtubules and actin filaments but their name derives from the fact that their filaments are intermediate in diameter between the other two. 1), 2) and 4) are untrue.

4.2 Since anti-tubulin antibodies were used we are studying microtubules. Treatment with colchicine leads to the disappearance of microtubules as shown at time 1. At time 3 a star-like cluster of fibres which has appeared at time 2 has become very long and extend throughout most of the cell. These fibres are presumably microtubules being initiated from the microtubule organising centre (the centriole). To check that the filaments were microtubules, you could treat the cells with colchicine which would cause their dissolution if they are microtubules.

4.3 48μm = 48000nm therefore 8000 G actin molecules would be needed to extend over a distance of 48μm. Each actin filament has two polymer strands and there are 25 actin filaments in the process. Thus there are 8000 x 2 x 25 = 400,000 G actin molecules in an average acrosome process. These could be assembled in about 5 seconds!

4.4 This is not an easy question to answer. Both treatments would be expected to show evidence for actin filaments. Actin filaments are labile but are stabilised by phalloidin, which binds along its length. Thus treatment 2) would be the best.

4.5

Statement	Actin filament	Microtubule	Intermediate filament
1) Is made up of a globular protein	x	x	
2) Is a fibrous protein			x
3) Capable of self assembly and disassembly	x	x	
4) ATP is incorporated into structure	x		
5) GTP is incorporated into structure		x	
6) Form of polymer			
a) 2 stranded helix	x		
b) hollow tube		x	

Statement	Actin filament	Microtubule	Intermediate filament
7) Diameter of filament			
a) 25nm		x	
b) 8nm	x		
c) 10nm			x
8) Requires accessory proteins to make lateral connection	x	x	
9) Monomer of more than one size			x
10) Is attached to the plasma membrane	x		
11) Involved in brush border/structure	x		
12) Involved in the structure of cilia and flagella		x	

4.6

1) b) and c)

2) c)

3) a), d) and f)

4.7

1) A glucose polymer linked α-1,4 would show a gentle helix because the twist at the 4 carbon link would be accentuated at the 1 carbon link. This molecule would not form straight chains but would gradually twist round and round. The significance of the β link orientation is that it counteracts the twist at carbon-4 and, therefore, produces strain chains which by hydrogen bond formation produce fibres with great strength in their longitudinal direction. If the cell walls of plants were made of α-1,4 linked compounds they would be no stronger than a thick rice pudding!

2) They are used as storage forms of carbohydrate. In plants such compounds as starch, are produced. In animals the normal product is glycogen.

4.8

1) c), e), f) and h)

2) b) and c)

3) b), d) and g)

4) b) and h)

Responses to Chapter 5 SAQs

5.1   A) light reaction.

B) ATP (or NADPH).

C) NADPH (or ATP).

D) dark reaction.

E) O_2.

F) sugar.

5.2   1) If we compare the absorption spectrum of the plant pigments and the action spectrum we see that only light which is absorbed can be used in photosynthesis.

If we compare the action spectrum with the two absorption spectra for chlorophylls a) and b) and the spectrum for carotenoids we can suggest that chlorophyll is the pigment which is involved in the process of photosynthesis.

2) The leaf appears green because it absorbs all wavelengths of light other than green. Thus green light passes from the leaf into our eyes, whereas all other colours are absorbed by the leaf and do not do this.

5.3   Transfers 2), 4) and 5) are energetically feasible.

5.4   If the permeability properties of the membrane could be modified so as to allow protons to diffuse through it. No proton gradient would occur and no ATP synthesis would result. You might have also suggested removing the ADP. This is an excellent suggestion. In carefully-prepared thyllakoids movement of protons through CF_o and CF_1 into the stroma is tightly coupled to the $ADP + P_i \rightarrow ATP$ reaction and will not occur in the absence of ADP. The chloroplast does not contain unlimited amounts of ADP and there is a requirement for ATP turnover, which refurnishes ADP, for the light reaction to continue.

5.5

Process	statements
1) cyclic photophosphorylation	b) involves only photosystem 1
	d) forms ATP only
2) non-cyclic photophosphorylation	a) forms NADPH, O_2 and ATP
	c) involves photosystems 1 and 2
	e) involves the oxidation of water
	f) involves P680

If you had difficulty with this question re-read section 5.3.5 - 5.3.9.

5.6   1) $3CO_2 + 9ATP + 6NADPH \rightarrow$ glyceraldehyde 3-phosphate $+ 9ADP + 8Pi + 6NADP^+$.

2) Non-cyclic photophosphorylation produces equal amounts of ATP and NADPH, the cyclic scheme produces only ATP. By controlling the ratio of operation of these two processes, plants can presumably modify the ratio of ATP : NADPH.

5.7   The concentration of 3-phosphoglyceric acid would increase and attract water into the chloroplast by osmosis. The chloroplast would swell and eventually burst. By turning the 3-phosphoglyceric acid into starch, which is insoluble, the chloroplast is able to maintain its solute concentration at a fairly constant level.

5.8   This is a somewhat open-ended question depending on how you define the term stable. The reduced components of the electron transport chain contain more energy than their oxidised counterparts and so does a thyllakoid that has a high internal H^+ concentration compared with one with a lower. You could argue that these are stable. However, they are not of very wide use. Thus the best answer here would be NADPH, a widely utilisable form of reducing power, and ATP, a widely utilisable source of energy.

5.9 NAD^+ is reduced to NADH during the production of pyruvate. During anaerobic metabolism NADH is oxidised to NAD^+ either by direct reaction with pyruvate forming lactic acid or by reducing acetaldehyde to ethanol, the acetaldehyde being formed from pyruvate. The NAD^+ is then available for the breakdown of more glucose. The reformation of NAD^+ in this way enables the continued breakdown of glucose with the production of 2ATP per glucose. Without this mechanism glucose metabolism would come to a stop.

5.10   Any situation in which you run out of oxygen in a tissue. This is a especially found in skeletal muscles when they are working hard (eg during running).

5.11   Because the uncharged form is lipophilic it will be able to diffuse through membranes. Charged molecules might 'absorb' the protons pumped through the inner membrane and transport them back again into the matrix. They would, therefore, discharge the formation of the proton gradient and no ATP would be formed.

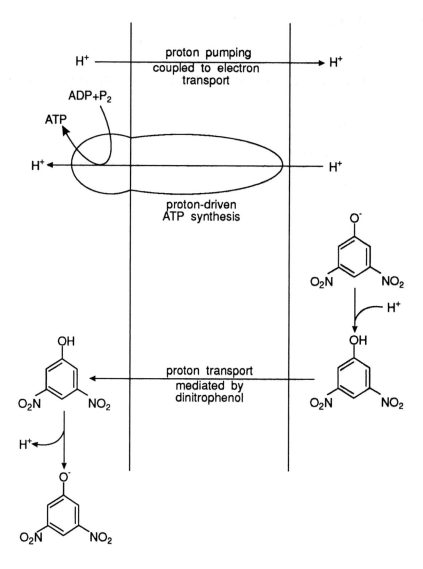

5.12 The figure of 38 ATP assumes that all the potential energy from NADH formed in glycolysis is released after its transferr to the mitochondria. NADH in the cytoplasm can be used in a number of reactions which oxidise it, so not all of it is oxidised in the mitochondria. Similarly if all the protons pumped out of the mitochondria were used to drive ATP formation the maximum number would be obtained. If any of the energy is used, however, to drive, for example, pyruvate or phosphate absorption, less ATP will be generated.

5.13 Oxygen is the final acceptor and the significance lies not only in the fact that much more ATP is generated. Because air contains approximately 20% oxygen and because oxygen can diffuse through membranes, using oxygen as the final hydrogen acceptor means using a compound which is present in virtually unlimited supply. Further the product, water, is not toxic. Ethanol, the product of acetaldehyde reduction, is toxic.

5.14   1) c), d), h)

2) b)

3) b), e) - f) may also be true although sucrose synthesis is not universal in heterotrophic cells.

4) b)

5) c), d), g)

6) a)

Responses to Chapter 6 SAQs

6.1 If the compounds were moving by active transport or by endocytosis no particular order would be expected. The data actually suggests that the compounds are moving by diffusion and the path could be the nuclear pore. Thus the smallest molecules move fastest and have easier access through the pore.

6.2 You may have come up with many different approaches. The most common method used is immunofluorescent staining. Immunofluorescent staining is a very sensitive method for detecting the nuclear lamina. For this, antibodies which react with nuclear lamina proteins are prepared and a fluorescence dye attached to them. If such antibodies are incubated with a thin section of a cell, they bind very strongly and specifically with the nuclear lamina proteins. We can visualise where they are by examining the cell for fluorescence using a light microscope.

6.3 Two loops round a histone core would be about 68nm, add 20nm for the linker and this gives 88nm of DNA arranged in a linear length of 31nm (11+20nm). Thus the nucleosome reduces the length by not quite two thirds (ie the reduction is almost three fold).

6.4 1) b) and c) - The histone cores are held together by DNA linkers.

2) b), e) and f)

3) b), d) and g)

4) a), b), e) and h)

6.5

		DNA	tRNA	rRNA	mRNA
1)	Is involved in translation		X	X	X
2)	Is an adapter molecule		X		
3)	Contains codons	X			X
4)	Contains an anticodon		X		
5)	Is in the form of a double helix	X			
6)	Is involved in transcription	X			X
7)	Contains the whole of the genetic code	X			
8)	Is made in the nucleus	X	X	X	X

Part 8 is a little bit difficult because we have not described where rRNA and tRNA is made. If you have worked out that all RNA must be made in the nucleus using the nucleotide sequence of DNA - well done.

6.6   1) 5' cap

2) poly-adenylic acid zone (poly A tail)

3) intron

4) messenger RNA

5) translation

6.7   There are many possible answers here but we would anticipate that you would include mRNA, tRNA and large and small ribosomal unit precursors.

6.8   This is quite a complex question. But a good answer should include the histone and non-histone proteins which are involved with DNA structure, the proteins that are incorporated with rRNA to form ribosomes and the three RNA polymerases which function in mRNA, rRNA and tRNA synthesis. We could also add the enzymes involved in RNA processing and the enzymes involved in DNA synthesis.

Responses to Chapter 7 SAQs

7.1

1) 124, one for each amino acid.

2) 372, ie 3x124, since each codon consists of 3 nucleotides.

7.2

The newly synthesised proteins would be expected to have alanine in place of cysteine. It is the tRNA molecule which recognises the anticodon and in the modified molecule the tRNA has not been changed and will still recognise the cysteine codon.

7.3

1) b), c), e) and f)

2) a), d), e) - since this amino acid occurs in proteins as well as acting as the initiation tRNA-amino acid) and i).

3) b), g) and i).

If you got these wrong it may be helpful to revise Section 7.2.4 by making a lage composite flow drawing of initiation elongation and termination.

7.4

Protein synthesis would occur until the ribosome moved along the mRNA to the point where the non-coding sequence was on the A site. There would not be a tRNA molecule whose anticodon fitted this codon so protein synthesis would stop at this point.

7.5

It suggests that poly-U was used as a mRNA and was translated to form polyphenylalanine. Since no other nucleotides were present the codon for phenylalanine must be UUU.

7.6

Our hint about removing all the spaces between the words in a sentence is we would not know where to start reading each word.

With a mRNA molecule consisting of GUUCACA, for example, possible codons could be GUU, UUC, UCA etc, depending on where the codon started. If the mRNA consisted of GUU alone, then clearly there is only one codon possible and the identification of the amino acid it codes for could be confirmed. This technique was instrumental in the elucidation of the code shown in Figure 7.7.

7.7

An ambiguous code could not guarantee to produce a protein with a specific amino acid sequence. Thus the primary structure of the protein would be altered and this would be expected to alter secondary and possibly tertiary structure. Thus the protein's function might be quite different, as seen by abnormal as opposed to normal haemoglobin.

7.8

1) phenylalanine-serine-histidine-tryptophan

2) the DNA would have the structure:

3' AAA AGC GTA ACC 5'

The way we worked this out was to remember the U is coded by A, and that C pairs with G. Thus for the mRNA 5' UUU UCG CAU UGG 3' the DNA must have the sequence 3' AAA AGC GTA ACC 5' (remember DNA is transcribed in an antiparallel manner). Remember that DNA is a double standard molecule. We have only shown one strand.

If you could not answer this question because you were unfamiliar with the base pairing arrangements of DNA and RNA, we could suggest you read the chapter on nucleic acids in the BIOTOL text 'The Molecular Fabric of Cells'.

7.9   Initially there would be silver grains over the ER. At a later time interval the grains would be over the Golgi apparatus and later still the grains would be over the secretory vesicles. Such experiments are called pulse-chase experiments.

7.10   1) a) small ribosome subunit; b) large ribosome subunit; c) leader sequence; d) membrane acceptor protein; e) ER lumen; f) removed leader sequence; g) amino end of growing peptide.

2) Ribophorin is embedded in the ER and the large ribosome subunit binds to it. It cannot be easily represented on Figure 7.10.

7.11   1) The protein would now contain several bulky oligosaccharides and it would presumably be very difficult for it to pass through the ER membrane into the lumen.

2) If every asn residue was glycosylated instead of a small number, this would drastically alter the folding of the protein and would almost certainly change its properties.

3) Asparagine, serine and threonine.

7.12   1) T

2) T

3) T

4) F, the protein does not become *attached* to the Golgi apparatus, they are entrapped within the Golgi lumen.

5) F, budding occurs at the trans face.

7.13   The nucleus has a double membrane. Remember, however, that the inner and outer membrane are continuous with each other at the nuclear pores. Material is considered to move from the outer to the inner membrane via the nuclear pores.

Responses to Chapter 8 SAQs

8.1
1) mitosis

2) cytokinesis

3) M-phase

4) interphase

5) S-phase

If you have managed to get all five correct answers you have clearly understood the material to date. If not, you should spend a little more time studying the previous sections.

8.2
1) True - It is of course critical that the DNA be replicated accurately and be distributed equally between the two daughter cells.

2) True - Remember the duration of the G_1-phase can vary from as little as a few hours, as in yeast, to as long as a year, as in the cells of the adult human liver.

3) False - This clearly cannot occur because it would mean that each daughter cell would not be able to receive a complete copy of the cell's DNA. You will see later precisely how the cell manages to ensure that mitosis will only occur after all the DNA has replicated.

8.3
1) False - You have to read this question carefully. It is certainly true that cells pass through a critical point in G_1 called start, but this does not trigger the onset of DNA replication. DNA replication occurs in the S-phase and is initiated by the S-phase activator.

2) False - Most proteins accumulate gradually throughout the cell cycle. There are, of course, exceptions and the one which we discussed is the synthesis of histones which is largely restricted to the S-phase.

3) True - Nothing really more to add.

4) False - The termination of DNA replication is accompanied by a great reduction in the level of the S-phase activator, or at least some essential component of it. The M-phase promoting factor is responsible for initiating chromosome condensation and the onset of mitosis. It does not terminate DNA replication.

8.4
1) You should have ticked possibility b). This was precisely the experiment that identified the S-phase activator. The G_1 nucleus is ready and able to initiate DNA replication, all it lacks is the appropriate signal which is clearly in abundance in the S-phase.

2) Possibility a) is correct here. The M-phase cell will contain the M-phase promoting factor which is a dominant control factor and will therefore initiate chromosome condensation in the G_1 nucleus. This will, of course, spell disaster for the G_1-phase cell because the premature condensation of its chromosomes will prevent any further progress through the cell cycle. During the normal cell cycle, elaborate controls determine that the appearance of the M-phase promoting factor is delayed until late in G_2.

8.5

a) This represents the metaphase. The nuclear membrane has broken down (note the presence of membrane vesicles), the sister chromatids have associated with the mitotic spindle and have aligned across the equator of the cell, the metaphase plate.

b) This is the telophase. The sister chromatids have moved to opposite poles and are beginning to de-condense. A new nuclear membrane is forming around each set of sister chromatids. Remember, when the cell divides (at cytokinesis) the sister chromatids will become the new cells' chromosomes.

c) This is prophase. What are the chief clues? a) The daughter centrosomes are just beginning to move apart and establish the microtubular network that will eventually create the mitotic spindle. b) The two sister chromatids, held together at their centromeres, are only just beginning to condense.

d) This is prometaphase. The daughter centrosomes have established the two spindle poles. There is an extensive microtubular network that comprises the mitotic spindle and the sister chromatids have become attached via their kinetochores. The chromatids, however, are still in active motion and have not yet collected along the metaphase plate.

This was not an easy question. If you successfully identified the four stages, well done. If not, you should go back and study this sequence again.

8.6

You should have drawn circles round kinetochores (which is where the chromatids become attached to the spindle), kinetochore microtubules (which are the microtubules of the spindle that attach to the kinetochore; shortening of these in anaphase A begins to move the sister chromatids to opposite poles of the spindle), and the polar microtubules (these are responsible for pushing the spindle poles apart during anaphase B, again affecting the separation of the sister chromatids).

The other structures are not involved in chromatid separation. Astral microtubules emanate from the spindle pole, but do not play a role in separation. The nuclear membrane has disappeared during prometaphase, long before sister chromatid separation. The contractile ring is involved in cytokinesis which occurs after sister chromatid separation.

8.7

1) 40 - The adult liver cell will be diploid, therefore it will possess 40 chromosomes, hence 40 DNA molecules.

2) 20 - A sperm will be haploid and thus contain 20 chromosomes.

3) 80 - The cell undergoing meiosis will be diploid (40 chromosomes), but these chromosomes will have replicated DNA during the interphase that precedes prophase I, therefore, the cell will contain 80 DNA molecules.

4) 40 - Each daughter cell will have a haploid number of chromosomes (ie 20) but each of these will carry sister chromatids (ie they have been replicated before the first meiotic division).

5) Each gamete (egg and sperm) contains a haploid number of chromosomes, each in a single copy. Thus each gamete has 20 molecules of DNA. On fertilisation, the egg and sperm nuclei fuse, to produce a normal diploid cell which will therefore contain 40 molecules of DNA.

Responses to Chapter 9 SAQs

9.1   1) a) - The cambial cells are totipotent because they are capable of generating all of the cells of the new plants. Totipotency is more common in plants than in animals.

2) c) - The colony forming unit cells give rise to more than one cell type but cannot produce all cell types, ie they are pluripotent, not totipotent.

3) b) - Red blood cells are committed to being one cell type. They cannot differentiate into other cell types. (ie they are irreversibly committed).

9.2   The epithelial cells in the intestine are continually exposed to mechanical as well as to chemical stress which means that there is a higher risk of damage. Pancreatic cells, on the other hand, are relatively free of such pressures. There is therefore a greater need for intestinal cells to be replenished because of the increased incidence of cell damage and death. This example serves to illustrate that not all cells need to be replenished at the same time or at the same rate.

9.3   You many have come up with one of several different answers here. We would argue that the single most important factor is the ability to replicate DNA. If cell renewal is to be achieved by simple cell division then the cell must be able to faithfully replicate its DNA in order to pass on duplicate copies to the daughter cells. Those cell types that have lost the ability to replicate DNA must be replaced by stem cells.

9.4   1) Simple cell division.

2) Stem cell differentiation.

3) Stem cell differentiation.

4) Stem cell differentiation.

5) Nerve cells cannot be replenished.

Cell renewal in liver, blood and skin is relative straightforward and the answers should have been no surprise to you. Muscle cell renewal perhaps posed a problem for you. The requirement for stem cell differentiation to replace muscle cells derives from the fact that the complex structure of the contents of the muscle cell renders a normal division process impossible. The nerve cell was the 'trick question' we hope you realised that damaged nerve cells are incapable of being replaced.

9.5   None of the cells show totipotency, since none of the cells shown are capable of proliferating and differentiating to form a complete organism. The haemopoietic cells, the common lyphoid progenitor cells and the common myeloid progenitor cells are all pluripotent since they can all produce more than one cell type. Note that some of the progeny of the haemopoietic cell have become more committed such that they can produce a more restricted range of cell types compared with the haemopoietic cell.

9.6   1) The most important function of desmosomes is to stabilise a tissue by means of mutual interaction between cells. Desmosomes therefore mainly occur in those tissues which are exposed continually to mechanical stress. There are several tissues which contain desmosomes. Three good examples are heart muscle and epithelial cells in the skin and the intestine.

2) The organisation of tight junctions in the intestinal epithelium is of such a nature that a perfect permeability barrier is created between the intestinal lumen and the intracellular space which is in direct contact with the capillary blood vessels. Nutrients therefore can only reach the blood by passing through epithelial cells. Tight junctions are also responsible for the maintenance of the bipolar nature of the epithelial cells in the intestine. They will prevent transport proteins in the membrane exchanging between the side of the cell that contacts the lumen of the intestine (the apical side) and the opposite side of the cell (the basolateral side) by lateral diffusion through the plasma membrane. This asymmetric distribution of transport systems is important for the specific transport of nutrients by the epithelial cells.

9.7   At one level there is no difference between proteoglycans and glycoproteins since in both cases there is a link between carbohydrate and protein. However, there is a difference when you consider the percentage of each in the final molecule; proteoglycans consist mainly of carbohydrate (90-95% on the basis of its weight) whereas the carbohydrate component of glycoproteins is considerably lower (up to a maximum of 60%, but some times considerably less than that).

9.8   1) hyaluronic acid.

2) tropocollagen.

3) elastin.

4) heparin.

If you did not get these right we would suggest you re-read the section on the intercellular matrix.

Appendix 1

Units of measurement

For historical reasons a number of different units of measurement have evolved. The literature reflects these different systems. In the 1960s many international scientific bodies recommended the standardisation of names and symbols and a universally accepted set of units. These units, SI units (Systeme Internationale de Unites) were based on the definition of: metre (m), kilogram (kg); second (s); ampare (A); mole (mol) and candela (cd). Although, in the intervening period, these units have been widely adopted, their adoption has not been universal. This is especially true in the biological sciences.

It is, therefore, necessary to know both the SI units and the older systems and to be able to interconvert between both sets.

The BIOTOL series of texts predominantly uses SI units. However, in areas of activity where their use is not common, other units have been used. Tables 1 and 2 below provides some alternative methods of expressing various physical quantities. Table 3 provides prefixes which are commonly used.

Mass (S1 unit: kg)	Length (S1 unit: m)	Volume (S1 unit: m^3)	Energy (S1 unit: $J = kg\ m^2\ s^{-2}$)
$g = 10^{-3}\,kg$	$cm = 10^{-2}\,m$	$l = dm^3 = 10^{-3}\,m^3$	$cal = 4.184\,J$
$mg = 10^{-3}\,g = 10^{-6}\,kg$	$Å = 10^{-10}\,m$	$dl = 100\,ml = 100\,cm^3$	$erg = 10^{-7}\,J$
$\mu g = 10^{-6}\,g = 10^{-9}\,kg$	$nm = 10^{-9}\,m = 10Å$	$ml = cm^3 = 10^{-6}\,m^3$	$eV = 1.602 \times 10^{-19}\,J$
	$pm = 10^{-12}\,m = 10^{-2}\,Å$	$\mu l = 10^{-3}\,cm^3$	

Table 1 Units for physical quantities

Concentration (SI units: $mol\ m^{-3}$)

a) $M = mol\ l^{-1} = mol\ dm^{-3} = 10^3\ mol\ m^{-3}$

b) $mg\,1^{-1} = \mu g\ cm^{-3} = ppm = 10^{-3}\ g\ dm^{-3}$

c) $\mu g\ g^{-1} = ppm = 10^{-6}\ g\ g^{-1}$

d) $ng\ cm^{-3} = 10^{-6}\ g\ dm^{-3}$

e) $ng\ dm^{-3} = pg\ cm^{-3}$

f) $pg\ g^{-1} = ppb = 10^{-12}\ g\ g^{-1}$

g) $mg\% = 10^{-2}\ g\ dm^{-3}$

h) $\mu g\% = 10^{-5}\ g\ dm^{-3}$

Table 2 Units for concentration

Fraction	Prefix	Symbol	Multiple	Prefix	Symbol
10^{-1}	deci	d	10	deka	da
10^{-2}	centi	c	10^2	hecto	h
10^{-3}	milli	m	10^3	kilo	k
10^{-6}	micro	μ	10^6	mega	M
10^{-9}	nano	n	10^9	giga	G
10^{-12}	pico	p	10^{12}	tera	T
10^{-15}	femto	f	10^{15}	peta	P
10^{-18}	atto	a	10^{18}	exa	E

Table 3 Prefixes for S1 units

Appendix 2

Chemical Nomenclature

Chemical nomenclature is quite a difficult issue especially in dealing with the complex chemicals of biological systems. To rigidly adhere to a strict systematic naming of compounds such as that of the International Union of Pure and Applied Chemistry (IUPAC) would lead to a cumbersome and overly complex text. BIOTOL has adopted a pragmatic approach by predominantly using the names or acronyms of chemicals most widely used in biologically-based activities. It is recognised however that there remains some potential for confusion amongst readers of different background. For example the simple structure CH_3COOH can be described as ethanoic acid or acetic acid depending on the environment or industry in which the compound is produced or used. To reduce such confusion, the BIOTOL series makes every effort to provide synonyms for compounds when they are first mentioned and to provide chemical structures where clarity and context demand.

Appendix 3

Abbreviations used for the common amino acids

Amino acid	Three-letter abbreviation	One-letter symbol
Alanine	Ala	A
Arginine	Arg	R
Asparagine	Asn	N
Aspartic acid	Asp	D
Asparagine or aspartic acid	Asx	B
Cysteine	Cys	C
Glutamine	Gln	Q
Glutamic acid	Glu	E
Glutamine or glutamic acid	Glx	Z
Glycine	Gly	G
Histidine	His	H
Isoleucine	Ile	I
Leucine	Leu	L
Lsyine	Lys	K
Methionine	Met	M
Phenylalanine	Phe	F
Proline	Pro	P
Serine	Ser	S
Threonine	Thr	T
Tryptophan	Trp	W
Tyrosine	Tyr	Y
Valine	Val	V

Index